Construction Contracts

Construction Contracts

Editor

Ambrish Kumar

scitus
academics

Construction Contracts

Edited by **Ambrish Kumar**

Printed in 2017

ISBN: 978-1-68117-349-8

Library of Congress Control Number: 2015939262

© 2016 by

SCITUS Academics LLC,
616, Corporate Way, Suite 2, 4766,
Valley Cottage, NY 10989

www.scitusacademics.com

Contents

Preface

Construction contract is simply mutual or legally binding agreement between two parties based on certain policies and conditions generally recorded in documentation form. The two parties involved are owner and contractor. The owner has full authority to decide what type of contract should be used for a specific facility to be constructed and to set forth the terms and conditions in a contractual agreement. Companies seeking to build an office building or industrial facility can document their agreement with the contractor through a construction contract. The contract will specify the scope of the work, which may include pre-construction services, bid analysis, and coordination with architects and subcontractors. The contract should include necessary payment and project completion schedules. In the event of a dispute, the parties may require the dispute to be resolved through arbitration, and any damages to be determined according to a liquidated damages clause.

Editor

Using Intelligent Techniques in Construction Project Cost Estimation: 10-Year Survey

Abdelrahman Osman Elfaki[1, 2] Saleh Alatawi[1, 2]
and Eyad Abushandi[1, 2]

[1]University of Tabuk, Tabuk 50060, Saudi Arabia
[2]Binladen Research Chair on Quality and Productivity Improvement in the Construction Industry, College of Engineering, University of Hail, Saudi Arabia

ABSTRACT

Cost estimation is the most important preliminary process in any construction project. Therefore, construction cost estimation has the lion's share of the research effort in construction management. In this paper, we have analysed and studied proposals for construction cost estimation for the last 10 years. To implement this survey, we have proposed and applied a methodology that consists of two parts. The first part concerns data collection, for which we have chosen special journals as sources for the surveyed proposals. The second part concerns the analysis of the proposals. To analyse each proposal, the following four questions have been set. Which intelligent technique is used? How have data been collected? How are the results validated? And which construction cost estimation factors have been used? From the results of this survey, two main contributions have been produced. The first contribution is the defining of the research gap in this area, which has not been fully covered by previous proposals of construction cost estimation. The second contribution of this survey is the proposal and highlighting of future directions for forthcoming proposals, aimed ultimately at finding the optimal construction cost estimation. Moreover, we consider the second part of our methodology as one of our contributions in this paper. This methodology has been proposed as a standard benchmark for construction cost estimation proposals.

INTRODUCTION

Information technology (IT) plays a crucial role in dealing with challenges in construction projects. Thomas et al. [1] have illustrated the importance of using IT to improve the performance of construction projects. The construction industry faces numerous complicated challenges that go beyond IT. These complicated challenges motivate the use of intelligent techniques to handle those challenges. For instance, intelligent techniques may be used to handle challenges such as (1) selecting the best-qualified prime contractor, (2) predicting project performance at different phases, or (3) estimating risk for cost overruns (running beyond a proper plan may lead to greater risks for many contractors). Recently, the civil engineering community has begun to consider Artificial Intelligence (AI) techniques as an optimal art for handling the above 3 fuzzy and ambiguous challenges [2]. The use of AI in the civil engineering sector has been introduced by Parmee [3], who proposes for AI to tackle problem areas characterised by uncertainty and poor definition.

Cost estimation is the most important preliminary process in any construction project. In the construction industry, cost estimation is the process of predicting the costs required to perform the work within the scope of the project [4]. Accurate cost estimation is crucial to ensure the successful completion of a construction project. Estimating construction cost is an example of a knowledge-intensive engineering task [5]; that is, it depends on the expertise of the human professional. In fact, engineers require several years to develop the necessary expertise to conduct the cost estimation process. The main problem here is that the engineers' expertise is often not documented or authenticated. Hence, this expertise is prone to subjectivity (i.e., defined to an extent by one's personal opinion). According to Shane et al. [6], accuracy and comprehensiveness in cost estimation are delicate issues and can be easily affected by many different parameters; in addition, each parameter must be properly addressed in order to maintain an acceptable level of accuracy during the process. Therefore, estimating construction cost to a fair degree of accuracy is mostly impossible to achieve manually.

On the other hand, inaccurate cost estimation leads to many problems, such as change order, construction delay [7], or even business bankruptcy in the worst scenarios. These two factors (i) the impossibility of conducting cost estimation manually and (ii) the effects of incorrect cost estimation thus encourage researchers and construction companies to investigate intelligent solutions to handle the problem of cost estimation.

This paper investigates and summarises the current use of intelligent solutions in the construction industry. In order to leverage the importance of intelligent solutions in project cost estimation, a list of state-of-the-art methods has been analysed, including machine-learning (ML), rule-based systems (RBS), evolutionary systems (ES), agent-based system (ABS), and hybrid systems (HS).

This paper has been organised as follows: we discuss our research methodology in Section 2. In Section 3, we explore and define the construction cost estimation factors that are used in this survey paper. In Section 4, intelligent techniques that are used in construction cost estimation are classified into five groups, and the main strengths and weaknesses of each group are defined. Each proposal is then analysed. Conclusions and future directions are presented in Section 5.

RESEARCH METHODOLOGY

The importance of cost estimation in the construction industry has been discussed in the previous section. However, there is no doubt that intelligent solutions may solve the dilemma of cost overruns, considering all affecting factors. In fact, there are a huge number of intelligent techniques available to deal with problems in construction management [8]. This motivates the researchers to carry out and analyse intelligent techniques with regard to tackling the construction cost estimation problem. This paper surveys the intelligent solutions employed over the last decade and identifies the directions for future development. This will help to provide more precise and in-depth analysis for the most recent proposals. The analytical process will highlight the research gap in this area. Furthermore, it will open a door for defining the available opportunities for future research.

This research has been divided into three parts, as shown in Figure 1. Firstly, we create a literature review database on the intelligent techniques that have been used in cost estimations over the last decade. In this step, specific journals have been selected based on their specialisation both in construction management and in information technology. These journals are Journal of Computing in Civil Engineering (http://ascelibrary.org/journal/jccee5), Journal of Construction Engineering and Management (http://ascelibrary.org/journal/jcemd4), Itcon (http://www.itcon.org/), Journal of Civil Engineering and Management (http://www.tandfonline.com/toc/tcem20/current#.VFIO8Wdh71U), and Automation in Construction (http://www.journals.elsevier.com/automation-in-construction/). Consequently, the collected papers have been classified based on their applied techniques. Secondly, we present an analysis and discussion of each intelligent technique to clarify its strengths and weaknesses. The strengths and weaknesses of specific intelligent techniques will be inherited by the cost estimation method based on that technique. Additionally, cost affecting factors have been established in order to carry out a specific benchmarking process.

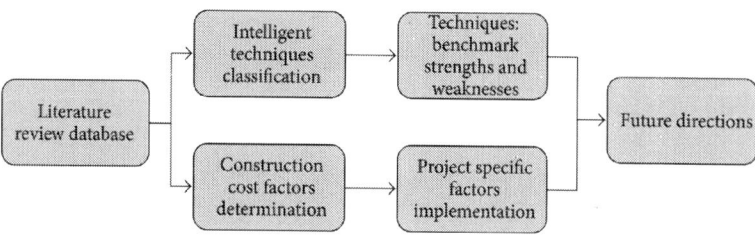

Figure 1: Flow chart illustrating the methodology.

Later, an intensive comparison of the surveyed construction projects' cost estimation methods, based on a proposed benchmark, has been conducted. To analyse each proposal, the research has focused on four points:

- The intelligent technique in use.
- How the proposal's data is collected.
- Validation of the proposed idea.
- The coverage of cost estimation factors.

We will now explain the steps from Figure 1 in detail. The first step is the creation of a literature database from the four journals mentioned earlier. We have used the words "construction cost" as a primary keyword; we have then selected only the proposals that involve the use of intelligent techniques. The second and third steps are parallel. In the second step, the intelligent techniques that have been found in selected proposals are classified based on the scientific concept of each one. The aim of this classification is to define the main features of each class. Step four describes the proposed benchmark, explained in detail in Table 2. This benchmark has been developed on the basis of construction cost factors selected in step five. Step six shows the last step and the main objective of this survey, which is the defining of future directions in the research area of construction cost estimation.

CONSTRUCTION COST FACTORS

According to Shane et al. [6], Oberlender and Trost [9], and Ahiaga-Dagbui and Smith [10], any construction cost estimation should be developed based on specific parameters such as type of project, material costs, likely design and scope changes, ground conditions, duration, size of project, type of client, and tendering method. Therefore, in this paper we have introduced these factors as a benchmark to compare between the cost estimation proposals.

There are various different factors that affect cost estimation in construction projects. These factors can be clustered into two distinct groups [11]: (i) estimator-specific factors and (ii) design and project-specific factors.

Estimator Specific Factors

The cost estimator can be one of the three parties: contractor, consultant, or owner. Based on the estimator's background and experience, cognitive biases or errors in cost estimates may occur accordingly [11]. In many cases, the cost estimator makes decisions based on the likely gains, or losses, of a venture and not necessarily based on the real outcome of the decision [12]. Moreover, the individual estimator may customise pricing based upon best local practices [13], which differ from country to country. For this reason, this paper will focus on design and project-specific factors.

Design and Project-Specific Factors

These factors include project size, type of project, ground conditions, type of client, material costs, likely design and scope changes, duration, tendering method [6, 9, 10], and contract type. In the following paragraphs, these factors are discussed in detail to explore their meanings and functions regarding cost estimation.

Project Size

There is a strong correlation between project size in square feet or metre and the number of labours. However, as the number of labours increases, the cost estimation of some items may have some biases and become more plausible (e.g., production rate estimation or tasks scheduling). There are many empirical studies on how project size can influence cost estimation (e.g., [14, 15]).

Type of Project

Undertaking particular types of projects requires a suitable choice of technology and equipment used, as well as suitable work methods. However, this can limit the choice of materials and size of crew to be employed; consequently, this will affect the project budget.

Project types can be classified under several different categories. In general, there are six major types of construction projects: (1) building construction, (2) special-purpose construction, (3) heavy construction,

(4) highway construction, (5) infrastructure construction, and (6) industrial construction.

Ground Conditions

Before tendering, ground condition should be one of the first concerns in any construction project. Without knowing the ground condition, the contractor should still presume to estimate the cost; however, if the assumption is not proper, this will lead to additional costs for bad ground condition.

Type of Client

As each construction project has its own client ideas, roles, and objectives, the characteristics of the contract and bidding behaviour are mainly affected by client type. There are seven different types of clients as classified by Drew et al. [16]:

- Government.
- Housing Authority.
- Other public sector clients.
- Large developers.
- Large industrial, commercial, and retailing organisations.
- Medium and small industrial, commercial, and retailing organisations.
- Other private sector clients.

Material Costs

The material selection-time, type of materials, and their availability in the local market all demonstrate a statistically significant impact on the cost estimation of construction projects. Materials can represent up to 70% of the project construction cost [17]; hence, any methods used to estimate the material cost accurately will reduce wastage and improve the major project's cost and time benefits. In addition, the quantity of material required must be accurately measured from the drawing and is not dependent on the crew performance or work method adopted [13].

However, this factor can vary dramatically and is highly dependent on the performance and work method used by the labours.

Likely Design and Scope Changes

Depending on their level of experience, the client retains more influence over the design and once on site during construction. Certain types of projects require the client to appoint a design firm (Figure 2) to design and inspect the project phases, in order to achieve the standards expected by the client.

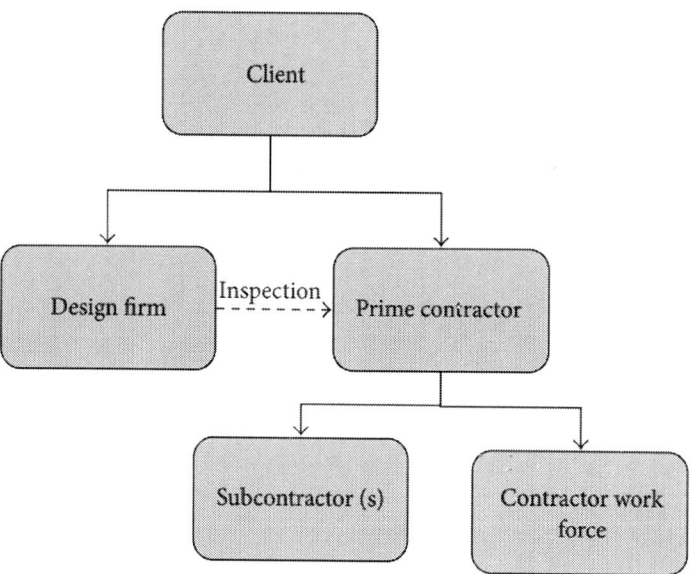

Figure 2: The client appoints a firm to design and inspect the project to meet certain specified (usually, oriented) requirements [50].

On the other hand, the right scope definition phase is highly important in the pre-project planning process. Poor scope definition is recognised by industry practitioners as one of the leading causes of project failure, as a high level of pre-project planning effort can result in around a 20% saving on total costs [18].

Duration

Research has indicated that there is a strong relationship between project cost and construction duration for different construction markets (e.g., [19, 20]). A relationship between completed construction cost and the time taken to complete a construction project was first mathematically established by Bromilow et al. [20]:where T is the duration of construction period, C is the final project cost, K is a constant value indicating the general level of duration performance, and B is a constant value describing how the duration performance is affected by the size of the construction project measured by its cost.

Figure 3 presents a duration-cost plane frame for small and medium infrastructure projects and identifies three main regions for project scheduling the boundaries of which are defined by general indexed project duration [21].

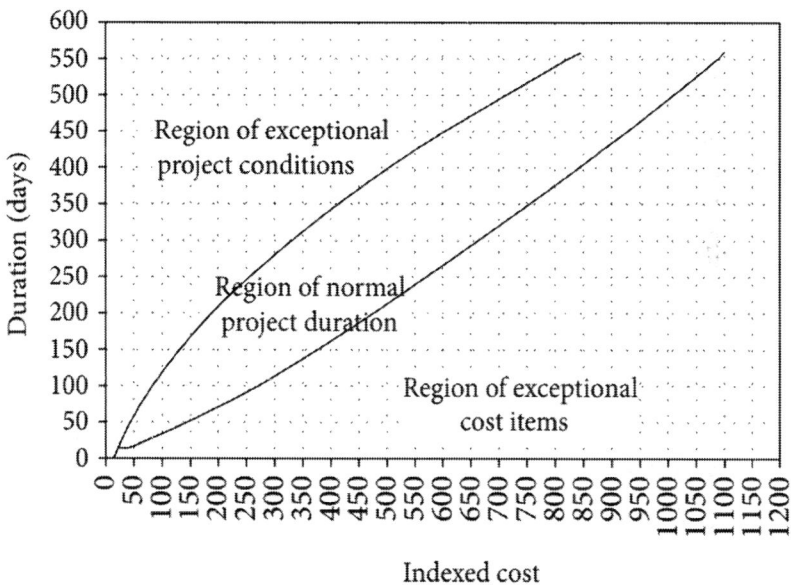

Figure 3: Modelled duration-cost envelope for policy decision support in small and medium projects [21].

Tendering Method

There are five tendering methods, including the following:

- Open Tendering. Contractors are invited to tender for a contract through local advertisements.
- Selective Tendering. Contractors are invited to tender on their proven record in relation to the type and size of contract and their own reliability.
- Negotiated Tendering. Cost reimbursement contract is a variation of this, which can be used when completion time is more important than cost.
- Two-Stage Tendering. It is used to bring in a contractor at the design stage, which is useful to advise the architect of any problems with the design of the building. Unit rates would be negotiated on the basis of the original tender.
- Serial Tendering. Tenders are invited from a selected list on the basis of a typical (notional) bill of quantities. The chosen contractor normally submits the lowest price and undertakes to enter into a series of contracts to carry out the work using the rates in the notional bill of quantities.

The selection of one of the above methods is basically intended to minimise any additional client risk. To achieve this goal, the client must balance four aspects:

- Client needs.
- Project cost.
- Completion time.
- Qualification of the tender to perform the job.

INTELLIGENT CONSTRUCTION PROJECT COST ESTIMATION METHODS

In this section, analysis of the surveyed intelligent construction cost estimation methods was conducted. These methods have been categorised into five groups, based on the intelligent technique that is used in each group: machine-learning (ML), rule-based systems (RBS), evolutionary systems (ES), agent-based system (ABS), and hybrid systems (HS).

At the first step, each group is explored to highlight their strengths and weaknesses. Subsequently, the methods are analysed in depth in terms of coverage of construction cost estimation techniques. In each proposal, four key questions have been highlighted for analysis. These questions are (1) which intelligent technique is used; (2) how the input datasets are collected; (3) how the proposed method is validated; and (4) which construction cost estimation factors are covered.

In the following subsections, firstly, the intelligent techniques employed are discussed, is the findings of which are considered as an answer to the first question. Secondly, each proposal is analysed individually, which answers question 2. The content of Table 1 illustrates the answer of question 3, while the content of Table 2 illustrates the answer of question 4.

Table 1: Comparison of proposals based on technique and validation

Proposal	Technique	Validation
Wilmot and Mei [24]	ML: ANN	Have not been mentioned
An et al. [26]	ML: SVM	Comparison with methods for assessing conceptual cost estimates
Petroutsatou et al. [23]	ML: ANN	By comparison with other models in literature
Jafarzadeh et al. [25]	ML: ANN	Have not been mentioned
Hola and Schabowicz [27]	ML: ANN	Have not been mentioned
Son et al. [28]	ML: SVM	Comparison with other techniques such as ANN and a decision tree (DT)
Cheng and Hoang [29]	ML: SVM	Have not been mentioned
Ji et al. [30]	KBS: case-based reasoning	Using case study

Choi et al. [31]	KBS: case-based reasoning	By comparison with previous conceptual cost estimation studies
K. J. Kim and K. Kim [32]	KBS: case-based reasoning	Have not been mentioned
Yildiz et al. [33]	KBS	By doing interview with experts
Lee et al. [34]	KBS: ontology	Comparison with other technique
Karakas et al. [41]	ABS: MAS	Interview with expert
Rojas and Mukherjee [42]	ABS: multiagent	Have not been mentioned
Kim [36]	KBS: case-based reasoning and analytical hierarchy process	Case study
de Albuquerque et al. [38]	ES: genetic algorithm	Have not been mentioned
Rogalska et al. [37]	ES: hybrid genetic evolutionary algorithm	By comparing the result with case studies from literature
Ghoddousi et al. [35]	ES: genetic algorithm	By comparing the result with case studies from literature
Afshar et al. [39]	ES: ant colony	By comparing the results with case studies in construction optimisation
Zhang and Ng [40]	ES: ant colony	By comparing the results with an academic benchmark
Kim et al. [43]	HS: statistics, CBR, and database	By comparing the result with case study from literature
Cheng et al. [44]	HS: SVM and DE	By comparing the result with other methods
Kim et al. [45]	HS: ANN and GA	Have not been mentioned
Yu and Skibniewski [46]	HS: ANN and fuzzy system	By using a case study of residential building construction projects in China

Williams and Gong [47]	HS: text mining, numerical data, and ensemble classifiers	Have not been mentioned
Cheng et al. [48]	HS: ANN, GA, and fuzzy system	Have not been mentioned
Zhang and Xing [49]	HS: fuzzy and particle swarm optimisation	Have not been mentioned

Table 2: Comparison of the proposals based on design and project-specific factors

Work	Project size	Project type	Ground conditions	Type of client	Likely design and scope changes	Contract type	Material costs	Duration	Tendering method
Wilmot and Mei [24]	Y	Y	Y	Y	Y	Y	Y	Y	N
An et al. [26]	Y	Y	Y	Y	N	N	N	Y	N
Petroutsatou et al. [23]	Y	Y	Y	Y	Y	Y	Y	N	N
Jafarzadeh et al. [25]	Y	Y	Y	Y	Y		N	N	N
Hola and Schabowicz [27]	Y	Y	Y	N	N	N	Y	Y	Y
Son et al. [28]	Y	Y	N	Y	Y	N	N	Y	N
Cheng and Hoang [29]	Y	Y	Y	N	N	N	N	Y	N
Ji et al. [30]	Y	Y	Y	N	Y	N	Y	N	N
Choi et al. [31]	Y	Y	Y	Y	N	N	N	N	N
K J. Kim and K 1Cim [32]	Y	Y	Y	Y	N	N	N	N	N
Yilcliz et al. [33]	Y	Y	N	N	Y	Y	N	Y	N
Lee et al. [34]	Y	Y	Y	N	Y	N	Y	N	N
1Caralcas et al. [41]	N	N	Y	Y	Y	N	Y	N	N
Rojas and Mukherjee [92]	Y	Y	Y	N	N	N	N	N	N
Kim [36]	Y	Y	Y	N	N	N	N	N	Y

de Albuquerque et al. [38]	Y	Y	Y	N	N		N		Y	N	N
Rogalslca et al. [37]	Y	Y	Y	N	N		N		N	N	N
Ghoddousi et al. [35]	Y	Y	Y	Y	N		N		N	N	N
Afshar et al. [39]	Y	Y	Y	N	N		Y		Y	Y	Y
Zhang and Ng [40]	Y	Y	Y	N	N		N		N	N	N
Kim et al. [43]	Y	Y	Y	N	N		N		Y	Y	N
Cheng et al. [94]	Y	Y	Y	N	N		N		Y	Y	N
Kim et al. [45]	Y	Y	Y	N	N		Y		Y	Y	N
Yu and Skibniewsld [96]	Y	Y	Y	Y	N		N		Y	Y	N
Williams and Gong [47]	Y	Y	Y	N	N		Y		Y	Y	Y
Cheng et al. [98]	Y	Y	Y	Y	N		N		Y	Y	N
Zhang and Xing [49]	Y	Y	Y	Y	N		Y		Y	Y	Y

Machine Learning (ML) Systems

ML systems have been defined as a construction of a system that can learn from data. In general, the main strengths of ML are (i) the ability to deal with uncertainty, (ii) the ability to work with incomplete data, and (iii) the ability to judge new cases based on acquired experiences from similar cases. On the other hand, the main weakness of ML is the lack of technical justification; that is, the causes beyond the decision are not available. This type of decision is called a black box decision. However, in the construction management, the artificial neural network (ANN) and the support vector machine (SVM) are the most common ML techniques. In the next paragraph, we analyse the construction cost estimation proposals that are based on ML.

One of the earliest papers to introduce the benefits and the implementation of ANN in the civil engineering community is published by Flood and Kartam [22]. This research has opened the door for many proposals that suggest ML as the preferred method to tackle various challenges in the construction industry. Petroutsatou et al. [23] introduced the ANN as a technique for early cost estimation of road tunnel construction. The data collection strategy of this research was based on structured questionnaires from different tunnel construction sites. The main drawback of this research was the ignoring of some of the construction cost factors (for more details, see Table 2). Wilmot and Mei [24] introduced an ANN model for highway construction costs. This research used the following factors as a base for cost estimation: price of labour, price of material, price of equipment, pay item quantity, contract duration, contract location, quarter in which the contract was let, annual bid volume, bid volume variance, number of plan changes, and changes in standards or specifications. The main contribution of this work was that it covered all required factors. Nevertheless, the validation of the proposed method and the data collection process used for training and testing the results were not fully presented. Jafarzadeh et al. [25] proposed the ANN method for predicting seismic retrofit construction costs. This study selected data from 158 earthquake-prone schools. The validation of this method is not clear. An et al. [26] proposed SVM for assessing conceptual cost estimates. Although this proposal is introduced as an assessment tool, still it might be considered as a cost estimation method. The method was developed on the basis of data from 62 completed building construction projects in Korea. Furthermore, Hola and Schabowicz [27] developed an ANN model for determining earthworks' execution times and costs. Basically, this model was developed on the basis of a database created from several studies that were carried out during large-scale earthwork operations on the construction site of one of the largest chemical plants in Central Europe. However, the validation of the presented results is not mentioned.

Son et al. [28] developed a hybrid prediction model that combines principal component analysis (PCA) with a support vector regression (SVR) predictive model for cost performance of commercial building projects. They used 64 related variables to define the pre-project planning stage. They developed their dataset based on information from 84 building projects in South Korea that had been completed within three years of the date at which the study was carried out. Questionnaires and interviews were used as a strategy for data collection. Cheng and Hoang [29] developed cost estimation at completion technique using least squares support vector machine. The data sets that are used in Cheng and Hoang [29] were collected from 13 reinforced concrete building projects executed between 2000 and 2007 by one construction Company headquartered in Taiwan.

Knowledge-Based Systems (KBS)

This category includes any technique that uses logical rules for deducing the required conclusions. The main strengths of KBS are (i) the ability to justify any result and (ii) uncomplicated methods (i.e., it is relatively easy to develop KBS). On the other hand, the limitations of KBS are (i) the difficulty of self-learning and (ii) time consumption during the rule acquisition process. Expert system and case-based reasoning are the common techniques used in KBS. The accuracy of case-based reasoning is highly dependent on the number of selected cases. Recently, KBS has been combined with other techniques to handle the limitation of the self-learning process. However, this mixture will be discussed in more detail in the section of this paper that deals with hybrid systems.

Ji et al. [30] proposed case-based reasoning to prepare strategic and conceptual estimations for construction budgeting. The data for this project were collected from 129 military barrack projects. Choi et al. [31] proposed a cost prediction model for public road planning.

The research data had been collected from a total of 207 real public road projects. Choi et al. [31] used rough-set theory to control the data collection and a genetic algorithm to optimise the rough-set model. Their work was classified as KBS since the authors implemented the case-based reasoning component in their cost estimation. K. J. Kim and K. Kim [32] developed a cost estimation model using CBR. This research overcomes the uncertainty in choosing the correct case by using a genetic algorithm. For this research, data were collected from 65 projects that constructed 585 bridges over a 5-year period. K. J. Kim and K. Kim [32] focused on construction of national bridges. However, it was not mentioned how the results were validated.

Yildiz et al. [33] developed a knowledge-based risk mapping tool to estimate costs for international construction projects. The required data and cost estimation parameters were collected from related literature. The validation process was performed in the form of expert interviews to get feedback on the developed tool. Lee et al. [34] proposed an ontological inference process for building cost estimation, by automating the process of searching for the most appropriate work items. Ghoddousi et al. [35] proposed a solution for determining total cost, time, and resources for construction projects; this was developed on the basis of a nondominated sorting genetic algorithm.

Kim [36] developed a cost estimation model based on case-based reasoning and analytical hierarchy process (AHP). In this project, data have been selected from literature and only 13 studies have been analyzed. Kim [36] developed his model based on data from highway construction projects. The validation has been conducted based on case study that contains data from 48 construction projects.

Evolutionary Systems (ES)

ES is a group of intelligent systems concerned with continuous optimisation with heuristics. As the results of ES are generated based on specific heuristics, they are very difficult to generalise, which is considered to be the main limitation of ES. The ability to solve complicated and uncertain problems is the main motivation for researchers to use ES. Evolutionary systems are used mainly as optimisation tools where there are many solutions; however, the ES algorithm assists in obtaining the correct solution.

Rogalska et al. [37] proposed a method based on genetic algorithm to deal with the problem of construction project scheduling. de Albuquerque et al. [38] developed a tool for estimating the cost of concrete structures. This tool is developed based on genetic algorithm. The cost has been estimated in all construction phases, such as manufacture, transport, and erection. Afshar et al. [39] developed a multicolony ant algorithm to solve the time/cost multiobjective optimisation problem. This method estimated both direct and indirect costs. Zhang and Ng [40] developed a Decision Support System (DSS) for cost estimation based on ant colony system. Zhang and Ng [40] used synthetic data to develop their DSS and they do validate their system by comparing it with a standard academic project. However, validation is done. Still Validation with real projects provide more accurate results.

Agent-Based System (ABS)

ABS has been considered as one of the main tracks in Artificial Intelligence, simulating the actions and interactions of autonomous agents with a view of assessing their effects on the system as a whole. In ABS, the generalisation of extracted results is the main challenge.

Karakas et al. [41] developed a multiagent system (MAS) that simulates the negotiation process between contractor and client regarding risk allocation and sharing of cost overruns in construction projects. This MAS was tested by interviewing eight professionals from the construction industry. In addition, Rojas and Mukherjee [42] developed a general multiagent simulation framework that can be used as an effective training environment. This framework could be used to estimate direct and indirect costs for construction projects.

Hybrid Systems (HS)

HS is defined as a collection of techniques used together to solve a specific problem. Usually, researchers use HS to overcome the techniques' individual limitations. Implementation of HS could represent a challenge, due to the unavailability of computational tools that could support its implementation. Furthermore, Kim et al. [43] proposed a hybrid conceptual cost estimating model for large mixed-use building projects. In this proposal, statistical analysis, CBR, and database methodologies were used together as a hybrid methodology. More recently, Cheng et al. [44] proposed a hybrid intelligence system for estimating construction cost. This hybrid system was developed based on support vector machine (SVM) and differential evolution (DE). In this proposal, data were collected across a number of public projects in Taiwan. Kim et al. [45] proposed hybrid models of ANN and GA for cost estimation of residential buildings, in order to predict preliminary cost estimates. In Kim et al.'s proposal, data were collected from residential buildings constructed in the years between 1997 and 2000 in Seoul, Korea. Yu and Skibniewski [46] proposed integrating a neurofuzzy system with conceptual cost estimation to discover cost-related knowledge from residential construction projects. The data used in this proposal was based on historical data from previous construction projects collected by the Ministry of Construction of PRC in the years between 1996 and 2002. Most recently, Williams and Gong [47] proposed text mining, numerical data and ensemble classifiers for estimating construction costs. Data used in this proposal were collected from 121 competitively bid highway projects. These data were collected from California Department of Transportation websites. Cheng et al. [48] proposed web-based conceptual cost estimates for construction projects, using an Evolutionary Fuzzy Neural Inference Model. Data were collected from 28 construction projects spanning the years from 1997 to 2001 in Taiwan. In this regard, Zhang and Xing [49] proposed a hybrid model for estimating construction costs, based on fuzzy and swarm optimisation. The data were collected from national bridge construction projects.

Table 1 shows the comparison of surveyed proposals, based on two issues. The first issue is the intelligent technique used in a proposal. The second issue is the type of validation that is used to prove the applicability of the proposal. Table 2 shows the comparison of surveyed proposals, based on design and project-specific factors used to estimate construction cost in each proposal. The letter "Y" means that this factor has been considered in this proposal, while the letter "N" means that this factor has not been considered. It is very obvious that there is no proposal that satisfies all the design and project-specific factors. On the other hand, in Table 1, there are some proposals that are provided without clear and scientific validation.

CONCLUSION AND FUTURE DIRECTIONS

In this paper, a survey and analysis were performed on different proposals in order to tackle the problem of developing construction cost estimation based on intelligent techniques. A scientific methodology has been designed to implement this survey. The method of the presented paper was based on two parts. The first part was concerned with a literature survey to examine the current state of intelligent solutions in the construction industry. Regarding this matter, we have chosen exclusively the journals that specialise in both information technology and construction management, within a time frame of ten years. In the research context, a ten-year period is sufficient to surround the directions of research in a specific area.

The second part was concerned with analysis of the proposals collected in the first part. Four key questions were selected to analyse each proposal. These questions are as follows:

- What is the intelligent technique used?
- How is the proposal's data collected?
- How is the proposed idea validated?
- What are the construction cost estimation factors used?

A justification of the four questions has been provided as follows:

- Defining the Intelligent Technique Used. This question is used to highlight the general strength and limitations of each proposal, which are reflected by the technique employed
- Defining Data Collection Method. This question is used to ensure the degree of accuracy. The degree of accuracy mainly depends on the collected data
- Defining the Validation of the Proposed Idea. This question is used to ensure the applicability of the proposed idea
- Defining the Commonly Used Cost Estimation Factors. This question is used to ensure the completeness of the proposal.

As mentioned in Section 3, there are two types of construction estimation factors: estimator-specific factors and design and project-specific factors. The first type, estimator-specific factors, depends on estimator expertise and skills and on lack of standardisation. The second type, design and project-specific features, is well defined and established in the civil engineering community. Due to the standardisation and stability of design and project-specific factors, this research paper considered only those factors mentioned in the designed methodology when applying the benchmark.

In conclusion, this paper provides two contributions to this area of knowledge: (1) an analysis of construction cost estimation proposals and (2) a standard survey methodology that can be used in any future surveys that deal with construction cost estimation.

According to the results of this research paper, the research gaps that have been deduced from this survey are as follows:

- There is a crucial necessity for a cost estimation method that covers all estimation factors from both types; that is, there is a need for one method that involves all "estimator specific" and "design and project-specific" factors. In Table 1, it is obvious that no proposal has a full row of "Y."

- There is a real need for a standard validation method which can be used to determine the accuracy level of a cost estimation proposal.

- There are many proposals that suffer from a lack of scientific justification for the results, that is, lack of describing how technically the results have been achieved.

Finally, future research directions are suggested for cost estimation in order to overcome the gaps that have been discussed. These directions are as follows:

- Providing cost estimation proposals that encourage the acquisition of human expertise: however, this releases the construction cost estimation from human dependability. Computerized expert systems are the better mechanism that might be used to replace human expertise. On another hand, knowledge management models and systems will assist in establishing computerized management systems that are free from the constraints of humanitarian. The main goal of knowledge management systems should be to capture and deal with estimator-specific factors. The first future direction is to encourage researchers and industry experts to adopt the direction of knowledge management systems in construction projects.

- Providing cost estimation proposals that are developed based on all "design and project-specific" factors: in Section 3, eight "design and project-specific" factors have been mentioned. The second future direction is to encourage researchers and industry experts to develop one integrated construction cost estimation system that works to achieve the all eight "design and project-specific" factors which have been mentioned.

- Providing a scientific justification for the cost estimation proposals based on real-world data: this will provide an explanation of how the estimates work and gives a justification on estimator's biases. Add the scientific justification for any proposal to increase the level of confidence in it. In addition, providing scientific justification assists in tracing the details of the cost estimation process which increase the level of transparency. Finally, providing scientific justification helps increase the maintainability.

- Providing a standard benchmark for determining the accuracy level of the construction cost estimation proposals: standard benchmarking leads to establishing a rule of thumb when other means of cost estimation are unavailable. This might be achieved by establishing a database containing information from previous projects. In addition, any future cost estimation models should consider this database "known value" to provide a useful benchmark for how accurately those models can estimate the cost. Using standard benchmark could help in classification, clustering, and ranking of cost estimation proposals.

The limitations of this research paper can be summarised in two points:

- Data was collected from specific journals only.
- The survey was limited to a ten-year period.

While this paper acknowledges these limitations, it is nevertheless able to provide valid answers on the current state of this area of research and to propose future directions.

ACKNOWLEDGMENTS

The present research work has been undertaken within the "Binladin Research Chair on Quality and Productivity Improvement in the

Construction Industry" at the University of Hail and funded by the Saudi Binladin Constructions Group.

REFERENCES

1. S. R. Thomas, S.-H. Lee, J. D. Spencer, R. L. Tucker, and R. E. Chapman, "Impacts of design/information technology on project outcomes," Journal of Construction Engineering and Management, vol. 130, no. 4, pp. 586–597, 2004.

2. H. G. Melhem, "Technical council for computing and information technology," Journal of Computing in Civil Engineering, vol. 22, no. 6, pp. 335–337, 2008.

3. I. C. Parmee, "Computational intelligence and civil engineering-perceived problems and possible solutions," in Towards a Vision for Information Technology in Civil Engineering, I. Flood, Ed., ASCE, Nashville, Tenn, USA, 2003.

4. L. Holm, J. E. Schaufelberger, D. Griffin, and T. Cole, Construction Cost Estimating: Process and Practices, Pearson Education, Upper Saddle River, NJ, USA, 2005.

5. S. Staub-French, M. Fischer, J. Kunz, and B. Paulson, "A generic feature-driven activity-based cost estimation process," Advanced Engineering Informatics, vol. 17, no. 1, pp. 23–39, 2003.

6. J. S. Shane, K. R. Molenaar, S. Anderson, and C. Schexnayder, "Construction project cost escalation factors," Journal of Management in Engineering, vol. 25, no. 4, pp. 221–229, 2009.

7. A. Albogamy, D. Scott, N. Dawood, and G. Bekr, "Addressing crucial risk factors in the middle east construction industries: a comparative study of Saudi Arabia and Jordan," Sustainable Building Conference Coventry University, West Midlands, UK, 2013.

8. T. Bulbul, C. J. Anumba, and J. Messner, "A system of systems approach to intelligent construction systems," in Proceedings of the ASCE International Workshop on Computing in Civil Engineering, pp. 22–32, Austin, Tex, USA, June 2009.

9. G. D. Oberlender and S. M. Trost, "Predicting accuracy of early cost estimates based on estimate quality," Journal of Construction Engineering and Management, vol. 127, no. 3, pp. 173–182, 2001.

10. D. D. Ahiaga-Dagbui and S. D. Smith, "Neural networks for modelling the final target cost of water projects," in Proceedings of the 28th Annual ARCOM Conference, S. D. Smith, Ed., pp. 307–316, Association of Researchers in Construction Management, Edinburgh, UK, September 2012.

11. B. Akinci and M. Fischer, "Factors affecting contractors' risk of cost overburden," Journal of Management in Engineering, vol. 14, no. 1, pp. 67–76, 1998.

12. A.-D. D. Dominic and S. D. Smith, "Rethinking construction cost overruns: cognition, learning and estimation," Journal of Financial Management of Property and Construction, vol. 19, no. 1, pp. 38–54, 2014.

13. N. Sinclair, P. Artin, and S. Mulford, "Construction cost data workbook," in Proceedings of the Conference on the International Comparison Program, World Bank, Washington, DC, USA, 2002.

14. A. Doyle and W. Hughes, "The influence of project complexity on estimating accuracy," in Proceedings of the 16th Annual ARCOM Conference, pp. 623–634, Glasgow Caledonian University, 2000.

15. I. Mahamid, "Early cost estimating for road construction projects using multiple regression techniques," Australasian Journal of Construction Economics and Building, vol. 11, no. 4, pp. 87–101, 2011.

16. D. Drew, M. Skitmore, and H. P. Lo, "The effect of client and type and size of construction work on a contractor's bidding 10 Advances in Civil Engineering strategy," Building and Environment, vol. 36, no. 3, pp. 393–406, 2001.

17. S. Donyavi and R. Flanagan, "The impact of effective material management on construction site performance for small and medium sized construction enterprises," in Proceedings of the 25th Annual Conference of the Association of Researchers in Construction Management (ARCOM '09), A. R. J. Dainty, Ed., pp. 11–20, Association of Researchers in Construction Management, Nottingham, UK, September 2009.

18. C. Cho and G. Edward, "Building project scope definition using project definition rating index," Journal of Architectural Engineering, vol. 7, no. 4, pp. 115–125, 2001.

19. A. P. Kaka and A. D. F. Price, "Relationship between value and duration of construction projects," Construction Management and Economics, vol. 9, no. 4, pp. 383–400, 1991.

20. F. J. Bromilow, M. F. Hinds, and N. F. Moody, The Time and Cost Performance of Building Contracts 1976–1986, Australian Institute of Quantity Surveyors, Sydney, Australia, 1988.

21. F. Edum-Fotwe, "Developing benchmarks for project schedule risk estimation," in System-Based Vision for Strategic and Creative Design, F. Bontempi, Ed., Swets & Zeitlinger, Lisse, The Netherlands, 2003.

22. I. Flood and N. Kartam, "Neural networks in civil engineering. I: principles and understanding," Journal of Computing in Civil Engineering, vol. 8, no. 2, pp. 131–148, 1994.

23. K. Petroutsatou, E. Georgopoulos, S. Lambropoulos, and J. P. Pantouvakis, "Early cost estimating of road tunnel construction using neural networks," Journal of Construction Engineering and Management, vol. 138, no. 6, pp. 679–687, 2012.

24. C. G.Wilmot and B. Mei, "Neural network modeling of highway construction costs," Journal of Construction Engineering and Management, vol. 131, no. 7, pp. 765–771, 2005.

25. R. Jafarzadeh, J. M. Ingham, S. Wilkinson, V. Gonzalez, and ´ A. A. Aghakouchak, "Application of artificial neural network methodology for predicting seismic retrofit construction costs," Journal of Construction Engineering and Management, vol. 140, no. 2, Article ID 04013044, 2014.

26. S.-H. An, U.-Y. Park, K.-I. Kang, M.-Y. Cho, and H.-H. Cho, "Application of support vector machines in assessing conceptual cost estimates," Journal of Computing in Civil Engineering, vol. 21, no. 4, pp. 259–264, 2007.

27. B. Hola and K. Schabowicz, "Estimation of earthworks execution time cost by means of artificial neural networks," Automation in Construction, vol. 19, no. 5, pp. 570–579, 2010.

28. H. Son, C. Kim, and C. Kim, "Hybrid principal component analysis and support vector machine model for predicting the cost performance of commercial building projects using preproject planning variables," Automation in Construction, vol. 27, pp. 60–66, 2012.

29. M.-Y. Cheng and N.-D. Hoang, "Interval estimation of construction cost at completion using least squares support vector machine," Journal of Civil Engineering and Management, vol. 20, no. 2, pp. 223–236, 2014.

30. S.-H. Ji, M. Park, and H.-S. Lee, "Case adaptation method of case-based reasoning for construction cost estimation in Korea," Journal of Construction Engineering and Management, vol. 138, no. 1, pp. 43–52, 2012.

31. S. Choi, D. Y. Kim, S. H. Han, and Y. H. Kwak, "Conceptual costprediction model for public road planning via rough set theory and case-based reasoning," Journal of Construction Engineering and Management, vol. 140, no. 1, Article ID 04013026, 2014.

32. K. J. Kim and K. Kim, "Preliminary cost estimation model using case-based reasoning and genetic algorithms," Journal of Computing in Civil Engineering, vol. 24, no. 6, pp. 499–505, 2010.

33. A. E. Yildiz, I. Dikmen, M. T. Birgonul, K. Ercoskun, and S. Alten, "A knowledge-based risk mapping tool for cost estimation of international construction projects," Automation in Construction, vol. 43, pp. 144–155, 2014.

34. S.-K. Lee, K.-R. Kim, and J.-H. Yu, "BIM and ontologybased approach for building cost estimation," Automation in Construction, vol. 41, pp. 96–105, 2014.

35. P. Ghoddousi, E. Eshtehardian, S. Jooybanpour, and A. Javanmardi, "Multi-mode resource-constrained discrete timecost-resource optimization in project scheduling using nondominated sorting genetic algorithm," Automation in Construction, vol. 30, pp. 216–227, 2013.

36. S. Kim, "Hybrid forecasting system based on case-based reasoning and analytic hierarchy process for cost estimation," Journal of Civil Engineering and Management, vol. 19, no. 1, pp. 86–96, 2013.

37. M. Rogalska, W. Bozejko, and Z. Hejducki, "Time/cost optimization using hybrid evolutionary algorithm in construction project scheduling," Automation in Construction, vol. 18, no. 1, pp. 24–31, 2008.

38.   A. T. de Albuquerque, M. K. El Debs, and A. M. C. Melo, "A cost optimization-based design of precast concrete floors using genetic algorithms," Automation in Construction, vol. 22, pp. 348–356, 2012.

39.   A. Afshar, A. K. Ziaraty, A. Kaveh, and F. Sharifi, "Nondominated archiving multicolony ant algorithm in time—cost trade-off optimization," Journal of Construction Engineering and Management, vol. 135, no. 7, pp. 668–674, 2009.

40.   Y. Zhang and S. T. Ng, "An ant colony system based decision support system for construction time-cost optimization," Journal of Civil Engineering and Management, vol. 18, no. 4, pp. 580–589, 2012.

41.   K. Karakas, I. Dikmen, and M. T. Birgonul, "Multiagent system to simulate risk-allocation and cost-sharing processes in construction projects," Journal of Computing in Civil Engineering, vol. 27, no. 3, pp. 307–319, 2013.

42.   E. M. Rojas and A. Mukherjee, "Multi-agent framework for general-purpose situational simulations in the construction management domain," Journal of Computing in Civil Engineering, vol. 20, no. 3, pp. 165–176, 2006.

43.   H.-J. Kim, Y.-C. Seo, and C.-T. Hyun, "A hybrid conceptual cost estimating model for large building projects," Automation in Construction, vol. 25, pp. 72–81, 2012.

44.   M.-Y. Cheng, N.-D. Hoang, and Y.-W. Wu, "Hybrid intelligence approach based on LS-SVM and Differential Evolution for construction cost index estimation: a Taiwan case study," Automation in Construction, vol. 35, pp. 306–313, 2013.

45.   G. H. Kim, D. S. Seo, and K. I. Kang, "Hybrid models of neural networks and genetic algorithms for predicting preliminary cost estimates," Journal of Computing in Civil Engineering, vol. 19, no. 2, pp. 208–211, 2005.

46.   W.-D. Yu and M. J. Skibniewski, "Integrating neurofuzzy system with conceptual cost estimation to discover cost-related knowledge from residential construction projects," Journal of Computing in Civil Engineering, vol. 24, no. 1, pp. 35–44, 2010.

47.   T. P. Williams and J. Gong, "Predicting construction cost overruns using text mining, numerical data and ensemble classifiers,"

Automation in Construction, vol. 43, pp. 23–29, 2014. Advances in Civil Engineering 11

48. M.-Y. Cheng, H.-C. Tsai, and W.-S. Hsieh, "Web-based conceptual cost estimates for construction projects using Evolutionary Fuzzy Neural Inference Model," Automation in Construction , vol. 18, no. 2, pp. 164–172, 2009.

49. H. Zhang and F. Xing, "Fuzzy-multi-objective particle swarm optimization for time-cost-quality tradeoff in construction," Automation in Construction, vol. 19, no. 8, pp. 1067–1075, 2010.

50. S. Nunnally, Construction Methods and Management, Prentice Hall, Englewood Cliffs, NJ, USA, 2007

Safety Climates in Construction Industry: Understanding the Role of Construction Sites and Workgroups

Sílvia Silva[1], Adriana Araújo[2], Dário Costa[3], and J. L. Meliá[4]

[1]ISCTE-IUL Instituto Universitário de Lisboa, Lisbon, Portugal
[2]ISCTE-IUL, Lisboa, Portugal
[3]Organizational and Social Psychology, ISCTE-IUL, Lisboa, Portugal
[4]Department of Methodology, University of Valencia, Valencia, Spain

ABSTRACT

Studies of safety climate in construction revealed a significant positive association between safety climate and various aspects of occupational health and safety. The mechanisms through which this impact operates are still unclear and safety climate is usually studied without considering the complexity of this industry (companies, worksites and groups). The aim of this research is to analyze to what extend there are differences between construction sites and to explore the relations between construction sites' safety climate and workers' safety response and to examine how this influence occur considering the workgroups. The safety climate was evaluated using a reduced version of the questionnaire that is a part of Battery HERC (Herramienta para evaluacion riesgos comportamentales). The data were collected in a Portuguese construction company (5 construction sites; including sub-contractors) comprising 213 workers. Differences between construction sites safety climate were found, suggesting the prevalence of safety sub-climates. The workgroup safety climate played a determinant role on workers' safety response in subcontracted workgroups and it is an important mechanism through which the principal contractor can influence subcontractors' safety response. Designers of prevention and training programs for accidents prevention should include specific contents in order to improve supervisory safety leadership and workgroup safety responses.

INTRODUCTION

According to Eurostat, more than one-in-four (26.1%) fatal accidents at work in the EU-27 in 2009 took place within the construction sector [1]. Work accidents are still a worrying phenomenon, with serious economic and social consequences. In Portugal, construction is one of highest risk industries. Even with recent reductions in incident rates, around 47% of workplace accidents registered in 2010 occurred in industries (26.6%) and construction (20.6%). Construction sector leads the total incidence rate, with 91,836 accidents per 100,000 workers, almost twice superior to the overall incidence rate. In 2012, in Portugal, 42 construction workers died [2].

Over the years, research community has tried to identify the factors associated with the accidents occurrence. Studies about safety culture and safety climate in construction focus on organizational and social factors, but safety climate have always been investigated separately at organization and subunit levels. Previous research also emphasizes that sub-climates for safety can exist within an organization. Some research on different groups within organizations has focused on comparing individuals who have not suffered an injury with those who have [3]. Glendon and Litherland [3] applied a modified version of the safety climate questionnaire [4] and found differences in the safety climate of job sub-groups on two of the factors: "Relationships" and "Safety Rules". Gillen and colleagues [5] found statistically significant differences between union and nonunion workers' responses regarding perceived safety climate. Cooper and Phillips [6] found different safety climate perceptions on departments within a company. Despite potential benefits of comparing sub-groups within an organization, few studies have evaluated different groups on a construction site basis. The aim of this research is to analyze to what extend there are differences on safety climate between construction sites and to explore the relations between construction sites' safety climate and workers' safety response.

SAFETY CLIMATE

Background

Organizational climate refers to shared perceptions among members of an organization with regard to aspects of the organizational environment that inform role behavior, that is, the extent to which certain facets of role behavior are rewarded and supported in any organization [7]. For instance Reichers & Schneider (1990: p. 22) emphasize the perceptions of organizational policies, practices and procedures, Since organizations have multiple goals and means of attaining goals, senior managers must develop policies and procedures for key organizational facets like customer service, product quality, and employees' safety [8]. Safety climate is a particular area of organizational climate that was introduced in the literature by Zohar [9] and is defined as individual perceptions of the policies, procedures and practices relating to safety in the workplace [10]. A range of factors has been identified as being important components of safety climate: management values (e.g. management concern for employee well-being), management and organizational practices (e.g. adequacy of training, provision of safety equipment, quality of safety management systems), communication and employee involvement in workplace health and safety [11]. The development of shared perceptions about the priority placed upon safety within the work environment is believed to inform workers' role behavior through expectations they form about how certain behaviors will be rewarded and supported in an organization [7,8]. During the past few decades, several researchers confirmed the effects of safety climate on employees' safety behaviors [6,11] and on accidents [12-17].

Recently, safety climate has been re-defined as a multilevel construct [7,13,16,17] that emphasized supervisors' safety practices behavior [13,18,19] and co-workers safety practices [16,17]. Meliá and colleagues [20], identified four main safety agents, organization, supervisors, co-workers and worker and five safety climate factors: Organizational safety response, supervisors' safety response, co-workers' safety response, worker's safety response and Perceived risk of accidents. This analysis is relevant because the organizational processes occur simultaneously at different hierarchical levels of an organization; policies and procedures are established at the organization level and are executed at the subunit level (supervisory practices). Policies define strategic goals and the procedures for their attainment, practices are related to the execution of policies by supervisory leaders across the organizational hierarchy [8]. For example, a supervisor who directs workers to disregard certain safety procedures whenever production falls behind schedule creates a distinction between company procedures and subunit practices, thus creating the potential for distinctive sub-climates within one organization.

In addition, the importance of co-workers as a safety climate agent has been reinforced and successfully tested. Recently, results from Brondino and colleagues [17] revealed that co-workers' safety climate had a stronger influence on safety behaviors, and in particular on safety participation, than supervisor's safety climate, at individual level as well at group level.

Construction Safety Climate

The study of the safety climate in construction began with Dedobbeleer and Beland [12] who tested the Brown and Holmes' three-factor safety climate model on construction workers, in nine non-residential construction sites. The results showed that safety was perceived as a joint responsibility of workers and management. Since then, many studies were conducted and revealed a significant positive association between safety climate and various aspects of occupational health and safety performance in the construction industry [5,21-23] and subsequent safety behaviors among Latino residential construction workers, with differences by trade being particularly important [24]. Teo & Feng [25], in their study about safety culture in construction sites, found that safety climate has an impact on the three dimensions of safety culture, psychological, situational/environmental and behavioral dimensions. According to Meliá and colleagues [20,26], safety climate can be analyzed from the point of view of the agent that performs the safety response in question, by identifying four main safety agents (organization, supervisors, co-workers and worker). In their study, results revealed that organizational safety response and supervisory safety response are strongly related, as are co-worker and worker safety response. In other studies, supervisors' safety response as perceived by workers has been considered a relevant part of safety climate models and therefore it has been included regularly in measures of safety climate, sometimes considering managers and supervisors together [6,27,28]. Lingard & Blismasa [29] tested a multi-level safety climate model in the Australian construction industry. Subcontracted workers' perceptions of the organizational safety response and supervisor safety response in their own organization and that of the principal contractor were measured using a safety climate survey and the results suggest that supervisors play an important role in shaping safety performance in subcontracted workgroups. The subcontracting is a main characteristic of construction industry and is determinant for the occupational health and safety performance. Subcontractor involvement is a core aspect of construction safety culture [30]. Construction subcontractors are often engaged in complex relationships both horizontally (i.e. when multiple subcontractors are engaged by a principal contractor) and vertically (i.e. in the case of pyramid of multilayered subcontracting). In this context, workers involved in subcontracted companies are not connected with the principal contractor and relatively isolated from their own company, which could affect the development and impact of the safety climate [26]. Moreover, considering that workers usually work in groups and have one supervisor it is important to see how much difference the workgroup can make in this specific context. These implications—subcontracting, construction sites and workgroups—and their impact on development of safety climate on the construction industry are still unclear.

The main goal of our research was to analyze the differences on safety climate among construction sites owned by the same principal company and to determine whether workgroup could play a mediation role between safety climate and worker safety response. We formulated the hypothesis as follows:

Hypothesis 1 [H1]: Differences on safety climate exists between construction sites.

Hypothesis 2 [H2]: Construction site safety climate is positive and significantly related with workers' safety response.

Hypothesis 2a [H2a]: The relationship between construction site safety climate and workers' safety response is mediated by workgroup safety climate, centered in supervisors.

Hypothesis 2b [H2b]: The relationship between construction site safety climate and workers' safety response is mediated by workgroup safety climate, centered in coworkers.

METHODOLOGY

Instrument

For the purpose of this study, the safety climate survey developed by Meliá was used. HERC is an instrument of safety climate that was developed and validated for the construction sector, on the Project "Elaboración y validación de una herramienta diagnostica estandarizada para la evaluación de los riesgos comportamentales y psicosociales ligados a siniestralidad en el sector de la contrucción". The version used comprises 3 parts. The questionnaire contains four scales to construction sites: construction site safety response (e.g. "In the construction site are conducted safety inspections to assess risks"), group safety climate, covering subscales related to co-workers and supervisors' safety responses (e.g. "My supervisor makes an effort to do his job safely") and workers' safety response (e.g. "When I do my work I follow safety instructions"). The questionnaire has 33 itens and the questions are answered in a six-point Likert scale (0 = never, 5 = continuously).

The questionnaire in this study was applied in one company responsible for five construction sites.

Cronbach's alpha was calculated for each scale to test internal consistency. The construction site safety climate scale revealed an adequate reliability (α = 0.75). The group safety response scales and workers' safety response scales also presented good reliability, supervisory α = 0.93 and co-workers, α = 0.90 and workers' safety response, α = 0.88.

Participants

Participants for this study were the construction workers on five construction sites performed by a Portuguese principal construction company. Overall, our sample covers 20 subcontracted companies, 57% of the total of subcontracted companies, and a total of 213 participants, approximately 65% of the total number of employees. The majority of the respondents were male (94.5%), 20.7% aged 26 - 30.

RESULTS

Table 1 shows means, standard deviations, and correlations among the variables used in the present study. The means show that respondents perceive high levels of workgroup safety climate, workers' safety response and low levels of construction site safety climate.

All variables are positive and significantly correlated. Construction site safety response and supervisors' safety response ($r = 0.51$, $p < 0.01$) are correlated and construction site safety climate and co-workers' safety response are correlated ($r = 0.39$, $p < 0.01$). Supervisors' safety response ($r = 0.63$, $p < 0.01$) and co-workers' safety response ($r = 0.49$, $p < 0.01$) have higher correlations with workers' safety response than other variables.

Comparison between Construction Sites

The mean values of the four types of safety responses, obtained in the five construction sites, were compared using an One-Way Anova test. Significant differences between construction sites were found for all the variables (F values range between 3.35 and 5.96; and p value between less than 0.001 to 0.01).

Table 1: Descriptive statistics and correlations

	M	SD	1	2	3
Construction site safety climate	2.77	0.99			
Supervisors' safety response	3.70	1.15	0.51**	OP	
Co-workers safety response	3.30	1.10	0.39**	0.52**	
Workers' safety response	4.03	0.88	0.27**	0.63**	0.49**

* $p < 0.05$; **$p < 0.01$.

Table 2 presents the results of the Tukey test (post hoc), to test mean differences between construction sites. As can be seen, in what concerns to construction site safety climate, construction site 1 (CS1) has lower means than all the others constructions sites and construction site 4 (CS4) has higher means. However, we only obtain significant differences between two construction sites (CS2 and CS4).

In what concerns to supervisory safety response, significant differences exists between one and other three construction sites, co-workers safety response presents significant differences between two construction sites and safety response presents significant differences between one and other two construction sites.

With regard to the shared perceptions within construction sites, it was found that only one construction site presented a high degree of consensus, (Rwg range between 0.03 and 0.74) which was expected considering the fact that there were employees from different companies.

Relation between Safety Climate and Workers' Safety Response and Mediating Role of Workgroup Safety Response

To test mediation a statistical procedure proposed by Baron and Kenny (1986) was applied and the Sobel test was calculated to check to what extent was or not the mediations significant. The assumptions were verified with correlations analysis.

Table 2: Variables mean values

	CS1	CS2	CS3	CS4	CS5
Construction site safety climate	1.9S	2.47a	2.29	3.02b	2.83
Supervisory safety response	3.03	4.29a	3.38b	3.67b	3.68b
Co-workers safety response	2.5a	3.83b	3.22	3.42	2.99a
Safety response	3.6	4.45a	3.69 b 3.91b		4.13

Means with different letters signify that groups are significantly different ($p < 0.05$).

Regarding the role of workgroup safety climate in the relationship between construction site safety climate and workers' safety response, (Table 3) results revealed that supervisors' safety response, is a complete mediator, since the relation between the predictor variable (construction safety climate) and the outcome variable (workers' safety response) is no longer significant with the introduction of the mediator variable (supervisors' safety response) in the model. Sobel test supports the complete mediation ($z = 6.78$, $p < 0.01$). This model explains 37% of total variance of workers' safety response.

Co-workers' safety response, is also a complete mediator in the relationship between construction site safety climate and workers' safety response (Table 4), since the relation between the predictor variable (construction site safety climate) and the outcome variable (workers' safety response) is no longer significant with the introduction of the mediator variable in the model (co-workers safety response). Sobel test supports that is a complete mediation ($z = 4.76$, $p < 0.01$). The model explains 23% of the total variance of workers' safety response.

To check if the same results would be obtained in the context of specific construction sites, the mediation hypothesis was also tested in the construction sites that had the biggest sample size namely CS2, CS4 and CS5. Overall, a similar mediation results pattern was also found. Namely, it was found a total mediation, supported by Sobel test, from supervisors and co-workers safety response on the relationship between construction site safety climate and workers safety response.

DISCUSSION

The aim of this research was to analyze to what extend there are differences between construction sites and to explore the relations between construction site safety climate and workers' safety response and to examine how this influence may occur. The present results showed that differences between construction sites exist.

The principal contribution of this study is the inclusion of a new agent in construction safety climate analysis—the construction site—which proved to be relevant for the analysis of safety climates.

Table 3: Supervisory safety climate mediation

	Criterion variables		
	Model 1	Model 2	Model 3
	Supervisory safety response	Workers' safety response	Workers' safety response
Predictor Variables	β	β	β
Construction site safety climate	0.51˙.	0.27*	—0.07

Supervisory safety response	-	-	0.65*
R²adjusted	0.26	0.06	0.37

* $p < 0.001$.

Table 4: Co-workers' safety climate mediation

	Criterion variables		
	Nlodel 1	Model 2	Model 3
	Co-workers safety response	Workers' safety response	Workers' safety response
Predictor Variables	β	β	β
Construction site safety climate	0.39*	0.27*	0.10
Co-workers safety response	-	-	0.44*
R² adjusted	0.15	0.06	0.23

* $p < 0.001$.

Indeed, construction site is an important, specific and complex part of the construction sector, which is so often referred to in an attempt to explain results in previous studies [20]. Construction subcontractors are often engaged in complex relationships and workers involved in subcontracted companies are not connected with the principal contractor and remain relatively isolated from their own company, which could conduct to sub-units (construction site) safety climate that should be predicted by construction companies.

Concerning to relations between construction site safety climate and workers safety performance, they are positively associated. The mediation role of workgroup, centered on supervisory safety climate and co-workers safety climate, was analyzed. Results shows, as previewed, that supervisory safety climate completely mediates the relationship between construction site safety climate and workers' safety response and that co-workers safety climate also completely mediates the relationship between construction site safety climate and workers' safety response. The results are in agreement with previous findings like the test of psychosocial model of workrelated accidents [16], that shows how safety climate influences workers' safety behavior through supervisors' and coworkers' safety responses and from Lingard, Cooke & Blismas [29], whose results suggest that supervisors play an important role in shaping safety performance in subcontracted workgroups.

There are some limitations of this study that should be noted. Self-reported data were used, and considering the nature of this context some degree of under-reporting, social desirability, and/or response bias may have occurred. Data were collected using only quantitative techniques. Notwithstanding these limitations, the present research contributes to the organizational safety literature by providing empirical evidence of an agent by which safety climate can be modified and supporting supervisor role in safety promotion.

CONCLUSIONS

This study focuses on the construction sites' specific safety climate that has been understudied. Moreover, previous studies on construction industry's mechanisms upon which safety climate has its impact were still unclear. The present study findings suggest the relevance of construction site safety climates, the prevalence of safety sub-climates and the importance of the workgroup safety climate on workers' safety response in subcontracted workgroups.

Future studies should continue to do deeper analysis that grasps better the nature of construction companies and construction sites. For instance, to analyze sub-contracted companies and it's workgroups operating in one specific construction site. It is important to see if there are differences between contracted companies and in what way the construction safety climate is different from the company safety climate and what ends up to be more determinant of workers' safety behaviors.

It will be also important to develop and test adequate intervention programs that can be applied to positively change the safety climate at all the levels as well as the individual's safety response.

Improving safety at work in the construction industry is still a challenge that requires multidisciplinary efforts for developing better prevention interventions.

REFERENCES

1. Eurostat, 2012. http://epp.eurostat.ec.europa.eu/statistics_ explained/index.php/Health_and_safety_at_work_statistics

2. ACT, 2013. http://www.act.gov.pt/(pt-PT)/CentroInformacao/ Estatistica/Paginas/default.aspx

3. A. Glendon and D. Litherland, "Safety Climate Factors, Group Differences and Safety Behavior in Road Construction," Safety Science, Vol. 39, No. 3, 2001, pp. 157- 188.http://dx.doi. org/10.1016/S0925-7535(01)00006-6

4. A. I. Glendon, N. A. Stanton and D. Harrison, "Factor Analyzing a Performance Shaping Concepts Questionnaire," In: S. A. Robertson, Ed., Contemporary Ergonomics, Ergonomics for All, Taylor and Francis, London, 1994, pp. 340-345.

5. M. Gillen, D. Baltz, M. Gassel, L. Kirsch and D. Vaccaro, "Perceived Safety Climate, Job Demands, and Coworker Support among Union and Nonunion Injured Construction Workers," Journal of Safety Research, Vol. 33, No. 1, 2002, pp. 33-51.http:// dx.doi.org/10.1016/S0022-4375(02)00002-6

6. M. D. Cooper and A. Phillips, "Exploratory Analysis of the Safety Climate and Safety Behavior Relationship," Journal of Safety Research, Vol. 35, No. 5, 2004, pp. 497-512.http://dx.doi. org/10.1016/j.jsr.2004.08.004

7. D. Zohar and G. Luria, "A Multilevel Model of Safety Climate: Cross-Level Relationships between Organization and Group-Level Climates," Journal of Applied Psychology, Vol. 90, No. 4, 2005, pp. 616-628. http://dx.doi.org/10.1037/0021-9010.90.4.616

8. D. Zohar, "Safety Climate and beyond: A Multi-Level Multi-Climate Framework," Safety Science, Vol. 46, No. 3, 2008, pp. 376-387.http://dx.doi.org/10.1016/j.ssci.2007.03.006

9.   D. Zohar, "Safety Climate in Industrial Organizations: Theoretical and Applied Implications," Journal of Applied Psychology, Vol. 65, No. 1, 1980, pp. 96-102.http://dx.doi.org/10.1037/0021-9010.65.1.96

10.   A. Neal and M. A. Griffin, "A Study of Lagged Relationships among Safety Climate, Safety Motivation, Safety Behaviour, and Accidents at the Individual and Group Levels," Journal of Applied Psychology, Vol. 91, No. 4, 2006, pp. 946-953.http://dx.doi.org/10.1037/0021-9010.91.4.946

11.   A. Neal, M. A. Griffin and P. M. Hart, "The Impact of Organizational Climate on Safety Climate and Individual Behavior," Safety Science, Vol. 39, 2000, pp. 157-188.

12.   N. Dedobbeleer and F. Beland, "A Safety Climate Measure for Construction Sites," Journal of Safety Research, Vol. 22, No. 2, 1991, pp. 97-103. http://dx.doi.org/10.1016/0022-4375(91)90017-P

13.   D. Zohar, "A Group-Level Model of Safety Climate: Testing the Effect of Group Climate on Micro-Accidents in Manufacturing Jobs," Journal of Applied Psychology, Vol. 85, No. 4, 2000, pp. 587-596. http://dx.doi.org/10.1037/0021-9010.85.4.587

14.   D. Zohar and G. Luria, "Climate as a Social-Cognitive Construction of Supervisory Safety Practices: Scripts as Proxy of Behavior Patterns," Journal of Applied Psychology, Vol. 89, No. 2, 2004, pp. 322-333. http://dx.doi.org/10.1037/0021-9010.89.2.322

15.   S. Silva, M. L. Lima and C. Baptista, "OSCI: An Organizational and Safety Climate Inventory," Safety Science, Vol. 42, No. 3, 2004, pp. 205-220.http://dx.doi.org/10.1016/S0925-7535(03)00043-2

16.   J. L. Meliá, "Un Modelo Causal Psicosocial de los Accidentes Laborales (A Psychosocial Causal Model of Work Related Accidents)," Anuario de Psicología, Vol. 29, No. 3, 1998, pp. 25-43.

17.   M. Brondino, S. Silva and M. Pasini, "Multilevel Approach to Organizational and Group Safety Performance: Co-Workers as the Missing Link," Safety Science, Vol. 50, No. 9, 2012, pp. 1847-1856. http://dx.doi.org/10.1016/j.ssci.2012.04.010

18.   D. Zohar and G. Luria, "The Use of Supervisory Practices as Leverage to Improve Safety Behavior: A Cross-Level Intervention

Model," Journal of Safety Research, Vol. 34, No. 5, 2003, pp. 567-577. http://dx.doi.org/10.1016/j.jsr.2003.05.006

19. J. L. Melià and A. Sesé, "Supervisor's Safety Response: A Multisample Confirmatory Factor Analysis," Psicothema, Vol. 19, No. 2, 2007, pp. 231-238.

20. J. L. Meliá, K. Mearns, S. Silva and M. L. Lima, "Safety Climate Responses and the Perceived Risk of Accidents in the Construction Industry," Safety Science, Vol. 46, No. 6, 2008, pp. 949-958. http://dx.doi.org/10.1016/j.ssci.2007.11.004

21. U. Varonen and M. Mattila, "The Safety Climate and Its Relationship to Safety Practices, Safety of the Work Environment and Occupational Accidents in Eight WoodProcessing Companies," Accident Analysis and Prevention, Vol. 32, No. 6, 2000, pp. 761-769. http://dx.doi.org/10.1016/S0001-4575(99)00129-3

22. S. Larsson, "Constructing Safety: Influence of Safety Climate and Psychological Climate on Safety Behaviour in Construction Industry," Chalmers University of Technology, Goteborg, 2005.

23. O. Siu, D. Phillips and T. Leung, "Safety Climate and Safety Performance among Construction Workers in Hong Kong. The Role of Psychological Strains as Mediators," Accident Analysis and Prevention, Vol. 36, No. 3, 2004, pp. 359-366. http://dx.doi.org/10.1016/S0001-4575(03)00016-2

24. T. A. Arcury, T. Mills, A. J. Marín, P. Summers, S. A. Quandt, J. Rushing, W. Lang and J. G. Grzywacz, "Work Safety Climate and Safety Practices among Immigrant Latino Residential Construction Workers," American Journal of Industrial Medicine, Vol. 55, No. 8, 2012, pp. 736-745. http://dx.doi.org/10.1002/ajim.22058

25. E. A.-L. Teo and Y. B. Feng, "The Role of Safety Climate in Predicting Safety Culture on Construction Sites," Architectural Science Review, Vol. 52, No. 1, 2009, pp. 5-16. http://dx.doi.org/10.3763/asre.2008.0037

26. J. L. Meliá, "An Integrative Multilevel Psychosocial view and measurement of Safety Climate," 11th European Congress on Work and Organizational Psychology, Lisbon, 2003.

27. S. J. Cox and A. J. T. Cheyne, "Assessing Safety Culture in Offshore Environments," Safety Science, Vol. 34, No. 1-3, 2000, pp. 111-129. http://dx.doi.org/10.1016/S0925-7535(00)00009-6

28. J. L. Melià and M. Becerril, "Safety Climate Dimensions from the 'Agent' Point of View," In: P. Mondelo, M. Mattila, W. Karwowski and A. Hale, Eds., Proceedings of the Fourth International Conference on Occupational Risk Prevention, Seville, 2006.

29. Helen Clare Lingard, Tracy Cooke and Nick Blismas, "Safety Climate in Conditions of Construction Subcontracting: A Multilevel Analysis," Construction Management and Economics, Vol. 28, No. 8, 2010, pp. 813-825.http://dx.doi.org/10.1080/01446190903480035

30. S. Mohamed, "Safety Climate in Construction Site Environments," Journal of Construction Engineering and Management, Vol. 128, No. 5, 2000, pp. 375-384.http://dx.doi.org/10.1061/(ASCE)0733-9364(2002)128:5(375)

Some Construction Methods of A-Optimum Chemical Balance Weighing Designs

Rashmi Awad and Shakti Banerjee

School of Statistics, Devi Ahilya University, Indore, India

ABSTRACT

Some new construction methods of the optimum chemical balance weighing designs and pairwise efficiency and variance balanced designs are proposed, which are based on the incidence matrices of the known symmetric balanced incomplete block designs. Also the conditions under which the constructed chemical balance weighing designs become A-optimal are also been given.

INTRODUCTION

Sir R. A. Fisher, a founder of modern concept of experimental designs gave the new ideas of designing in his first book Design of Experiment in the year 1935. Fisher's work was continued by others; see [1] - [4]

. The necessary and sufficient condition for a general block design to be variance balanced and efficiency balanced was given in the literature [5] - [8] . The concept of repeated blocks was introduced by Van Lint; see [9] . Further some potential applications of the balanced incomplete block designs with repeated blocks were presented in the literature [10] - [13] .

Another important concept which we discuss in this paper is weighing designs. The concept of weighing design was originally given by Yates and formulated as a weighing problem by Hotelling and the condition of attaining the lower bound by each of the variance of the estimated weights was given by him; see [14] [15] . In the latter developments, attention has been made in the direction of obtaining optimum weighing designs. Prominent work has been done by many researchers in this field; see [16] - [20] . In recent years, the new methods of constructing the optimum chemical balance weighing designs and a lower bound for the variance of each of the estimated weights from this chemical balance weighing design were obtained and a necessary and sufficient condition for this lower bound to be attained was proposed in the literature; see [21] - [24] . The constructions were based on the incidence matrices of balanced incomplete block designs, balanced bipartite block designs, ternary balanced block designs and group divisible designs.

Awad et al. [25] [26] gave the construction methods of obtaining optimum chemical balance weighing designs using the incidence matrices of symmetric balanced incomplete block designs and some pairwise balanced designs were also been obtained which were efficiency as well as variance balanced. In that series we now propose another new construction methods of obtaining optimum chemical balance weighing designs using the incidence matrices of symmetric balanced incomplete block designs and some more pairwise efficiency as well as variance balanced designs are proposed. Also we present the conditions under which the chemical balance weighing designs constructed by new construction methods leading to the A-optimal designs.

Let us consider   treatments arranged in b blocks, such that the jth block contains j k experimental units and the ith treatment appears r_i times in the entire design, i = 1,2, , ; j = 1,2, ,b . For any block design there exist a incidence matrix N = $[n_{ij}]$ of order   × b , where n^{ij} denotes the number of experiment units in the jth block getting the ith treatment.

When $n^{ij} = 1$ or 0 $\forall i$ and j, the design is said to be binary. Otherwise it is said to be nonbinary. In this paper we consider binary block designs only.

The following additional notations are used $\underline{k} = [k_1 k_2 ....k_b]$ is the column vector of block sizes, $\underline{r} = [r_1 r_2 ....rb_v]'$ is the column vector of treatment replication, $k_{b \times b} = diag[k_1 k_2 ....k_b], R_{v \times v} = diag[r_1 r_2 ...r_v], \Sigma r_i = \Sigma k_j = n$ is the total number of experimental units, with this $N1_b = \underline{r}$ and $N'1_v = \underline{k}$ Where 1_a is a$\times 1$ vector of ones.

An equi-replicate, equi-block sized, incomplete design, which is also balanced in the sense given above is called balanced incomplete block design, which is an arrangement of V symbols (treatments) into b sets (blocks) each containing $k(k < v)$ distinct symbols, such that any pair of distinct symbols occurs in exactly λ sets. Then it is easy to see that each treatment occurs in $(> \lambda)$ r sets. v, b , r , k , λ , are called parameters of the BIBD and the parameters satisfies the relations v r = bk , r $(k -1) = \lambda$ (v −1) and $b \geq$ v (Fisher's Inequality). A BIB design is said to be symmetric if b =v and $r = k$. In this case incidence matrix is a square matrix i.e. $N'=N$. In case of symmetric balanced incomplete block design any two blocks have λ treatments in common.

Though there have been balanced designs in various senses (see [6] [27]). We will consider a balanced design of the following type.

A block design is called variance balanced if and only if:

- It permits the estimation of all normalized treatment contrasts with the same variance (see [7]).

- If the information matrix for treatment effects $C = R - N K_{-1} N$ satisfies $C = \mu[I_v - (1/v)1_v 1'_v]$.

where μ is the unique nonzero eigen value of the matrix C with the multiplicity (v-1), I_v is the $v \times v$ identity matrix.

A block design is called efficiency balanced if

- Every contrast of treatment effects is estimated through the design with the same efficiency factor.

- $M_o = R^{-1}NK^{-1}N' - (1/n)1_v r' = \psi(I_v (1/n)1_v r')$; see [2] , and since $M_o S = \psi S$, where ψ is the unique non zero eigen value of M_o with multiplicity (v-1). For the EB block design N, the information matrix C is given as $C = (1-\psi)(R-(1/n)rr')$; see [28].

A block design is said to be pairwise balanced if $\sum_{j=1}^{b} n_{ij} n_{i'j} = \wedge$ (a constant) for all i, I';, $i \neq i'$ and a pairwise balanced block design is said to be binary if $n_{ij} = 0$ or 1 only, for all i; j; and it has parameters , b , r , k , Λ ($= \lambda$, say) , [in this case, when $\underline{r} = r1_v$ and $\underline{k} = k1_b$, it is a BIB design with parameters , b , r , k ,].

Weighing designs consists of n groupings of the p objects and suppose we want to determine the individual weights of p objects. We can fit the results into the general linear model

$$\underline{Y} = X\underline{w} + \underline{e}$$

(1)

where $\underline{Y}$ is an $n \times 1$ random column vector of the observed weights, $\underline{w}$ is the $p \times 1$ column vector repre- senting the unknown weights of objects and $\underline{e}$ is an $n \times 1$ random column vector of errors such that $E(\underline{e}) = 0_n$ and $E(\underline{e}\underline{e}') = \sigma^2 I_n$, $X = (x_{ij})$, $(i = 1,2,..., n; j = 1,2,..., p)$ is a $n \times p$ matrix of known quantities. The elements of matrix X take the values as

$$x_{ij} = \begin{cases} +1 & \text{if the } j_{th} \text{ object is placed in the left pan in the } i_{th} \text{ weighing,} \\ -1 & \text{if the } j_{th} \text{ object is placed in the right pan in the } i_{th} \text{ weighing} \\ 0 & \text{if the } j_{th} \text{ object is not weighted in the } i_{th} \text{ weighing} \end{cases}$$

The normal equations estimating w are of the form

$$X'X\hat{\underline{w}} = X'\underline{Y}$$

(2)

where $\hat{\underline{w}}$ is the vector of the weights estimated by the least squares method.

The matrix X is called the design matrix. A weighing design is said to be singular or nonsingular, depending on whether the matrix X'X is singular or nonsingular, respectively. It is obvious that the matrix X'X is nonsingular if and only if the matrix X is of full column rank (= p). Now, if X is of full rank, that is, when X'X is nonsingular, the least squares estimate of $\underline{w}$ is given by

$$\underline{\hat{w}} = \left(X'X \right)^{-1} X'\underline{Y}$$

(3)

and the variance-covariance matrix of $\underline{\hat{w}}$ is

$$\mathrm{Var}\left(\underline{\hat{w}} \right) = \sigma^2 \left(X'X \right)^{-1}$$

(4)

When the objects are placed on two pans in a chemical balance, we shall call the weighings two pan weighing and the design is known as two pan design or chemical balance weighing design. In chemical balance weighing design, the elements of design matrix $X = (x_{ij})$ takes the values as +1 if the j_{th} object is placed in the left pan in the i_{th} weighing, +1 if the j_{th} object is placed in the right pan in the i_{th} weighing and 0 if the j_{th} object is not weighted in the i_{th} weighing.

Hotelling has shown that if n weighing operations are to determine the weights of p=n objects, the minimum attainable variance for each of the estimated weights in this case is σ^2 / n and proved the theorem that each of the variance of the estimated weights attains the minimum if and only if $X'X = nI_p$ (see [14]).

VARIANCE LIMIT OF ESTIMATED WEIGHTS

Let X be an $n \times p$ matrix of rank p of a chemical balance weighing design and let m_j be the number of times in which j_{th} object is weighed, j=1,2...,p (i.e. the m_j be the number of elements equal to −1 and 1 in j_{th} column of matrix X). Then Ceranka et al. (see [21]) proved the following theorem:

Theorem 2.1: For any $n \times p$ matrix X, of a nonsingular chemical balance weighing design, in which maximum number of elements equal to -1 and 1 in columns is equal to m, where $m = \max\{m_1, m_2..., m_p\}$.

Then each of the variances of the estimated weights attains the minimum if and only if

$$X'X = mI_p$$

(5)

Also a nonsingular chemical balance weighing design is said to be optimal for the estimating individual weights of objects if the variances of their estimators attain the lower bound given by,

$$\mathrm{Var}\left(\hat{w}\right) = \frac{\sigma^2}{m}, \quad j = 1, 2, \cdots, p$$

(6)

In SBIB design $D(v,r,\lambda)$; the block intersection between any two blocks is constant i.e. λ. Using this concept Banerjee (see [29]) proved the following results;

Proposition 2.2: Existence of SBIB design $D(v,r,\lambda)$; implies

the existence of a BIB design D'with parameters $v'=v$, $b' = 2\binom{v}{2}$,

$r' = r(v-1)$, $k' = k$, $\lambda' = 2\binom{k}{2}$.

Proposition 2.3: Existence of SBIB design $D(v,r,\lambda)$; implies the

existence of a BIB design D'with parameters $v'=v$, $b' = \lambda\binom{v}{2}$, $r' = r\binom{r}{2}$

$, \ k' = k, \lambda' = \lambda \binom{k}{2}.$

CONSTRUCTION OF DESIGN MATRIX: METHOD I

In SBIB design D with the parameters v b = , r k = , ; fix the j_{th} block (j b = 1,2, , L) . Corresponding to the j_{th} fixed block, give negative sign to all the common treatments of remaining (b −1) blocks. Then eliminate that fixed block. Thus matrix N_*1 of design D_*1 is obtained. Now doing the same procedure for all the remaining (b −1) blocks, the incidence matrix $N*_1$ of the new design $D*_1$ so formed is the matrix having the elements 1, −1 and 0; given as follows

$$N_{*1} = \left[N_1 \ \vdots \ N_2 \ \vdots \cdots \vdots \ N_v \right]$$

(7)

Then combining the incidence matrix N of SBIB design repeated s-times with N^{*1} we get the matrix X of a chemical balance weighing design as

$$X = \left[N_{*1} \ \vdots \ \overbrace{N \quad \cdots \quad N}^{s\text{-times}} \right]'$$

(8)

Under the present construction scheme, we have $n = 2\binom{v}{2} + sb$ and p=v. Thus the each column of X will contain $\rho_1 = r(b-r) + sr$ elements equal to 1, $\rho_2 = r(b-r)$ elements equal to -1 and $n - \rho_1 - \rho_2$ elements equal to zero. Clearly such a design implies that each object is weighted $m = \rho_1 = \rho_2 = r(b-1) + sr$ times in $n = 2\binom{v}{2} + sb$ weighing operations.

Lemma 3.1: A design given by X of the form (8) is non singular if and only if $r(b-r) \neq (-k)(4+s)$.

Proof: For the design matrix X given by (8), we have

$$XX = \left[\{r(v-1)+sr\} - \{k(k-1)-4\lambda(k-\lambda)-s\lambda\}\right]I_v + \{k(k-1)-4\lambda(k-\lambda)+s\lambda\}J_{vv}$$
$$\Rightarrow XX = \left[\{r(b-r)+(k-\lambda)(4\lambda+s)\}\right]I_v + \{k(k-1)-\lambda\{4(k-\lambda)-s\}\}J_{vv} \tag{9}$$

and

$$|XX| = \left[r(b-1+s)+(v-1)\{k(k-1)-\lambda\{4(k-\lambda)-s\}\}\right] \times \left[r\{b-1-s\}-\{k(k-1)-\lambda\{4(k-\lambda)-s\}\}\right]^{v-1} \tag{10}$$

the determinant (10) is equal to zero if and only if $r(b-1+s) = k(k-1) - (4(k-)-s) \Rightarrow r(b-r) = (-k)(4+s)$ or $r(b-1+s) = (1-)\{k(k-1) - (4(k-)-s)\}$ but $r(b-1+s) + (-1)\{k(k-1) - (4(k-)-s)\}$ is positive and then det $(XX) = 0$ if and only if $r(b-r) = (-k)(4+s)$. So the lemma is prove

Theorem 3.2: The non-singular chemical balance weighing design with matrix X given by (8) is optimal if and only if

$$k(k-1) = \lambda\left[4(k-\lambda)-s\right] \tag{11}$$

Proof. From the conditions (5) and (9) it follows that a chemical balance weighing design is optimal if and only if the condition (11) holds. Hence the theorem.

If the chemical balance weighing design given by matrix X of the form (8) is optimal then

$$\mathrm{Var}\left(\hat{w}_j\right) = \frac{\sigma^2}{r(b-1+s)}; \quad j = 1, 2, \cdots, p$$

Example 3.3: Consider a SBIB design with parameters $v = b = 7$, $r = k = 4$, $= 2$; whose blocks are given by (3,5,6,7), (1,4,6,7), (1,2,5,7), (1,2,3,6), (2,3,4,7), (1,3,4,5), (2,4,5,6).

Theorem 3.2: yields a design matrix X of optimum chemical balance weighing design as

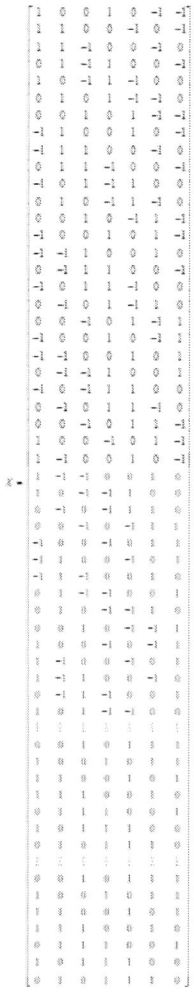

Clearly such a design implies that each object is weighted $m = 32$ times in $n = 56$ weighing operations and Var $(\hat{w}_j)$ 2 32 $w = \sigma$ for each $j = 1, 2, ... , 7$.

Corollary 3.4: If the SBIB design exists with parameters $v = b = N$, $r = k = (N \pm d)\,2$, $\lambda = (N \pm 2d + 1)\,4$; then the design matrix N_{*1} so formed using above method is optimum chemical balance weighing design.

Corollary 3.5: If in the design D_{*1}; -1 is replaced by zero then the new design D_{**1} so formed is a BIB design with parameters $B = 2\begin{pmatrix} v \\ 2 \end{pmatrix}$,

$R = r(b - r)$, $K = k - \lambda$, $\wedge = 2\begin{pmatrix} k - \lambda \\ 2 \end{pmatrix}$. Then the structure

$$N^{*1} = \left[N_{**1} \overset{\overbrace{s\text{-times}}}{\vdots N \quad \cdots \quad N} \right]$$

(12)

form a pairwise VB and EB design D^{*1} with parameters

$$v^{*1} = V, \quad b^{*1} = B + sb, \quad r^{*2} = R + sr,$$
$$k_1^{*1} = k - \lambda, \quad k_2^{*1} = k, \quad \lambda^{*1} = \Lambda + s\lambda,$$
$$\mu^{*1} = v\left[(k - \lambda - 1) + \frac{s\lambda}{k} \right] \text{ and } \psi^{*1} = 1 - \frac{V}{r^{*1}}\left[(k - \lambda - 1) + \frac{s\lambda}{k} \right].$$

CONSTRUCTION OF DESIGN MATRIX: METHOD II

In SBIB design D with the parameters $v = b$, $r = k$, λ; consider the λ blocks containing any pair of treatments say (θ, ψ). Now rearranging the λ -blocks corresponding to the pair (θ, ψ) and giving the negative sign to the treatments θ and ψ both; the matrix N_1 of design D_1 is obtained.

Now doing the same procedure for all the $\binom{v}{2}$ sets of blocks, the incidence matrix N_{*2} of the new design D_{*2} so formed is the matrix having the elements 1, -1 and 0; given as follows

$$N_{*2} = \left[N_1 \vdots N_2 \vdots \cdots \vdots N_{\binom{v}{2}} \right]$$

(13)

Then combining the incidence matrix N of SBIB design repeated S-times with N_{*2} we get the matrix X of a chemical balance weighing design as

$$X = \left[N_{*2} \vdots \overbrace{N \quad \cdots \quad N}^{s\text{-times}} \right]'$$

(14)

Under the present construction scheme, we have $n = \lambda \binom{v}{2} + sb$ and p=v. Thus the each column of X will contain $\rho_1 = 3\binom{r}{3} + sr$ elements equal to 1, $\rho_2 = r(k-1)$ elements equal to 1 and $n - \rho_1 - \rho_2$ elements equal to zero. Clearly such a design implies that each object is weighted

$n - \rho_1 - \rho_2 = r\binom{k}{r} + sr$ times in $n = \lambda \binom{v}{2} + sb$ weighing operations.

Lemma 4.1: A design given by X of the form (14) is nonsingular if and only if $\lambda/2[vr - (k-4)^2 \neq s(\lambda - r)$.

Proof. For the design matrix X given by (14), we have

$$X'X = \left[\left\{ r\binom{k}{2} + sr \right\} - \left\{ \frac{\lambda}{2} \left[(k-4)^2 - k \right] + s\lambda \right\} \right] I_v$$

$$+ \left\{ \frac{\lambda}{2} \left[(k-4)^2 - k \right] + s\lambda \right\} J_{vv}$$

$$\Rightarrow X'X = \left[\frac{\lambda}{2} \left[vr - (k-4)^2 \right] + s(r-\lambda) \right] I_v$$

$$+ \left\{ \frac{\lambda}{2} \left[(k-4)^2 - k \right] + s\lambda \right\} J_{vv}$$

(15)

and

$$|X'X| = \left[\left\{ r\binom{k}{2} + sr \right\} + (v-1) \left\{ \frac{\lambda}{2} \left[(k-4)^2 - k \right] + s\lambda \right\} \right]$$

$$\times \left[\left\{ r\binom{r}{2} + sr \right\} - \left\{ \frac{\lambda}{2} \left[(k-4)^2 - k \right] + s\lambda \right\} \right]^{v-1}$$

(16)

the determinant (16) is equal to zero if and only if

$$r\binom{k}{2} + sr$$

$$= \frac{\lambda}{2} \left[(k-4)^2 - k \right] + s\lambda$$

$$\Rightarrow \frac{\lambda}{2} \left[vr - (k-4)^2 \right]$$

$$= s(\lambda - r)$$

Or $\qquad r\binom{k}{2} + sr = (1-v)\left\{ \dfrac{\lambda}{2}\left[(k-4)^2 - k\right] + s\lambda \right\}$ $\qquad$ but

$r\binom{k}{2} + sr = (v-1)\left\{ \dfrac{\lambda}{2}\left[(k-4)^2 - k\right] + s\lambda \right\}$ is positive and then det(X'X) = 0 if and only if $\lambda / 2[\mathrm{vr} - (k-4)^2] = s(\lambda - r)$. So the lemma is proved.

Theorem 4.2: The non-singular chemical balance weighing design with matrix X given by (8) is optimal if and only if

$$(k-4)^2 = [k - 2s]$$

$$(17)$$

Proof: From the conditions (5) and (15) it follows that a chemical balance weighing design is optimal if and only if the condition (17) holds. Hence the theorem.

If the chemical balance weighing design given by matrix X of the form (14) is optimal then

$$\mathrm{Var}\left(\hat{w}_j\right) = \dfrac{\sigma^2}{r\left[\binom{k}{2} + s\right]}; \quad j = 1, 2, \cdots, p$$

Example 4.3: Consider a SBIB design with parameters v b = = 7, or k = = 3, = 1; whose blocks are given by (1,2,4), (2,3,5), (3,4,6), (4,5,7), (1,5,6), (2,6,7), (1,3,7).

Theorem 4.2 yields a design matrix X of optimum chemical balance weighing design as

$$X = \begin{bmatrix}
-1 & -1 & 0 & 1 & 0 & 0 & 0 \\
-1 & 0 & -1 & 0 & 0 & 0 & 1 \\
-1 & 1 & 0 & -1 & 0 & 0 & 0 \\
-1 & 0 & 0 & 0 & -1 & 1 & 0 \\
-1 & 0 & 0 & 0 & 1 & -1 & 0 \\
-1 & 0 & 1 & 0 & 0 & 0 & -1 \\
0 & -1 & -1 & 0 & 1 & 0 & 0 \\
1 & -1 & 0 & -1 & 0 & 0 & 0 \\
0 & -1 & 1 & 0 & -1 & 0 & 0 \\
0 & -1 & 0 & 0 & 0 & -1 & 1 \\
0 & -1 & 0 & 0 & 0 & 1 & -1 \\
0 & 0 & -1 & -1 & 0 & 1 & 0 \\
0 & 1 & -1 & 0 & -1 & 0 & 0 \\
0 & 0 & -1 & 1 & 0 & -1 & 0 \\
1 & 0 & -1 & 0 & 0 & 0 & -1 \\
0 & 0 & 0 & -1 & -1 & 0 & 1 \\
0 & 0 & 1 & -1 & 0 & -1 & 0 \\
0 & 0 & 0 & -1 & 1 & 0 & -1 \\
1 & 0 & 0 & 0 & -1 & -1 & 0 \\
0 & 0 & 0 & 1 & -1 & 0 & -1 \\
0 & 1 & 0 & 0 & 0 & -1 & -1 \\
\vdots & \vdots & \vdots & \vdots & \vdots & \vdots & \vdots \\
1 & 1 & 0 & 1 & 0 & 0 & 0 \\
0 & 1 & 1 & 0 & 1 & 0 & 0 \\
0 & 0 & 1 & 1 & 0 & 1 & 0 \\
0 & 0 & 0 & 1 & 1 & 0 & 1 \\
1 & 0 & 0 & 0 & 1 & 1 & 0 \\
0 & 1 & 0 & 0 & 0 & 1 & 1 \\
1 & 0 & 1 & 0 & 0 & 0 & 1
\end{bmatrix}$$

Clearly such a design implies that each object is weighted m=12 times in n=28 weighing operations and $Var(\hat{w}_j) = \sigma^2 / 12$ for each j = 1, 2…7.

Corollary 4.4: If the SBIB design exists with block size r ≤ 6 and ≤ 5; then the design matrix X so formed using above method II is optimum chemical balance weighing design

Corollary 4.5: If the SBIB design exists with parameters (v, v -1, v -2); then the design matrix X given by (14) is optimum chemical balance weighing design if and only if v ≤ 7.

Corollary 4.6. If in the design $D_{*2;.} - 1$ is replaced by zero then the new design D_{**2} so formed is a BIB design with parameters

$V=v, \quad B = \lambda \binom{v}{2}; \quad R = 3\binom{r}{3}; \quad k=k-2, \quad \wedge = \lambda / 2\left[(k-2)(k-3)\right].$ Then the structure

$$N^{*2} = \left[N_{**2} \vdots \overbrace{N \quad \cdots \quad N}^{s\text{-times}} \right]$$

(18)

form a pairwise VB and EB design D^{*2} with parameters

$$v^{*2} = V, \quad b^{*2} = B + sb, \quad r^{*2} = R + sr, \quad k_1^{*2} = k - 2, \quad k_2^{*2} = k, \quad \lambda^{*2} = \Lambda + s\lambda,$$

$$\mu^{*2} = v\lambda \left[\frac{(k-3)}{2} + \frac{s}{k} \right] \quad \text{and} \quad \psi^{*2} = 1 - \frac{v\lambda}{r^{*2}} \left[\frac{(k-3)}{2} + \frac{s}{k} \right]$$

A-OPTIMALITY OF CHEMICAL BAL-ANCE WEIGHING DESIGN

Some problems related to the optimality of chemical balance weighing designs were considered in the literature; see [17] [30] [31]. Wong and Masaro [32] [33] gave the lower bound for $\text{tr}\left[(X'X)^{-1}\right]$ and some construction methods of the A-optimal chemical balance weighing designs. Let X be a $n \times p$ design matrix of a chemical balance weighing design.

Then the following results from

Ceranka et al. [34] give the lower bound for $\text{tr}\left[(X'X)^{-1}\right]$.

Theorem 5.1: For any nonsingular chemical balance weighing design with the design matrix $X = \left(x_{ij}\right)$ we have

$$\text{tr}\left[(X'X)^{-1}\right] \geq \frac{p^2}{q \cdot n}$$

(19)

where $q = \max(q_1, q_2, \ldots, q_n); q_i = \sum_{j=1}^{p} x^2_{ij}, i = 1, 2 \ldots, n, ,$

The case when q=p; we get the inequality given in Wong and Masaro [32].

Definition 5.2. Any nonsingular chemical balance weighing design with the design matrix X=(x_{ij}) is said to be A-optimal if

$$\text{tr}\left[\left(X'X\right)^{-1}\right] = \frac{p^2}{q \cdot n}$$

(20)

Theorem 5.3: Any nonsingular chemical balance weighing design with the design matrix X=(x_{ij}) is A-optimal if and only if

$$X'X = \frac{q \cdot n}{p} I_p$$

(21)

CHECKING THE A-OPTIMALITY IN METHODS I AND II

For the construction Method I of chemical balance weighing design; the Lemma 3.1 proven above gave the necessary condition for the design matrix X of the form (8) to be non-singular.

Theorem 6.1: The non-singular chemical balance weighing design with matrix X given by (8) is A-optimal if and only if

$$k(k-1) = \lambda\left[4(k-\lambda) - s\right]$$

(22)

and

$$r(b-r)+(k-\lambda)(4\lambda+s) = \frac{q\left[v(v-1)+sb\right]}{v}$$

(23)

Proof: For the design matrix X given in (8) we have

$$X'X = \left[\{r(v-1)+sr\}-\{k(k-1)-4\lambda(k-\lambda)+s\lambda\}\right]I_v + \{k(k-1)-4\lambda(k-\lambda)+s\lambda\}J_{vv}$$
$$\Rightarrow X'X = \left[\{r(b-r)+(k-\lambda)(4\lambda+s)\}\right]I_v + \left[k(k-1)-\lambda\{4(k-\lambda)-s\}\right]J_{vv}$$

and

Comparing these two equalities we get

$$k(k-1) = \lambda\left[4(k-\lambda)-s\right]$$

and

$$r(v-1)+sr-\{k(k-1)-4\lambda(k-\lambda)+s\lambda\} = \frac{q\left[v(v-1)+sb\right]}{v} \Rightarrow r(b-r)+(k-\lambda)(4\lambda+s) = \frac{q\left[v(v-1)+sb\right]}{v}$$

If (22) is satisfied then we get the condition (23) from the last equation. Hence the theorem.

For the construction Method II of chemical balance weighing design; the Lemma 4.1 proven above gave the necessary condition for the design matrix X of the form (14) to be non-singular.

Theorem 6.2. The non-singular chemical balance weighing design with matrix X given by (14) is A-op- timal if and only if

$$(k-4)^2 = (k-2s)$$

(24)

and

$$\frac{\lambda}{2}\left[vr-(k-4)^2\right]+s(r-\lambda)=\frac{q\left[\lambda\binom{v}{2}+sb\right]}{v} \tag{25}$$

Proof. For the design matrix X given in (14) we have

$$X'X=\left[\left\{r\binom{k}{2}+sr\right\}-\left\{\frac{\lambda}{2}\left[(k-4)^2-k\right]+s\lambda\right\}\right]I_v+\left\{\frac{\lambda}{2}\left[(k-4)^2-k\right]+s\lambda\right\}J_{vv}$$

$$\Rightarrow X'X=\left[\frac{\lambda}{2}\left[vr-(k-4)^2\right]+s(r-\lambda)\right]I_v+\left\{\frac{\lambda}{2}\left[(k-4)^2-k\right]+s\lambda\right\}J_{vv}$$

and

$$X'X=\frac{q\left[\lambda\binom{v}{2}+sb\right]}{v}I_v$$

Comparing these two equalities we get

$$(k-4)^2=(k-2s)$$

and

$$\left\{r\binom{k}{2}+sr\right\}-\left\{\frac{\lambda}{2}\left[(k-4)^2-k\right]+s\lambda\right\}=\frac{q\left[\lambda\binom{v}{2}+sb\right]}{v}\Rightarrow\frac{\lambda}{2}\left[vr-(k-4)^2\right]+s(r-\lambda)=\frac{q\left[\lambda\binom{v}{2}+sb\right]}{v}$$

If (24) is satisfied then we get the condition (25) from the last equation. Hence the theorem.

DISCUSSION

The following Table 1 and Table 2 provide the list of pairwise variance and efficiency balanced block designs for Methods I and II respectively, which can be obtained by using certain known SBIB designs.

CONCLUSION

It is well known that pairwise balanced designs are not always efficiency as well as variance balanced. But in this research we have significantly shown that the proposed pairwise balanced designs are efficiency as well as variance balanced. Further there is a scope to propose different methods of construction to obtain the optimum chemical balance weighing designs and pairwise variance and efficiency balanced block designs, which will ful- fill the optimality criteria by means of efficiency. In this research paper we also gave the conditions under which the constructed chemical balance weighing designs lead to A-optimal designs. The only limitation of this research is that the obtained pairwise balanced designs all have large number of replications.

Table 1: For method I

S. No.	v^{*1}	b^{*1}	r^{*1}	k_1^{*1}	k_2^{*1}	λ_2^{*1}	μ^{*1}	φ^{*1}	Reference No.**
1	7	56	18	2	3	4	11.6667	0.3519	R (10), MH (1)
2	7	56	20	2	4	6	14.0000	0.3000	R (11)
3	11	132	40	3	5	10	30.8000	0.2300	R (29), MH (5)
4	11	132	42	3	6	12	33.0000	0.2143	R (30)

Table 2: For method II

S. No.	v^{*2}	b^{*2}	r^{*2}	k_1^{*2}	k_2^{*2}	λ^{*2}	μ^{*2}	ψ^{*2}	Reference No**.
1	4	16	6	1	3	0	2.6667	0.5556	R (2)
2	5	40	20	2	4	3	15.0000	0.25	R (4)
3	6	72	40	3	5	12	33.6	0.16	R (8)
4	7	28	6	1	3	0	2.3333	0.61111	R(10), MH (1)
5	7	56	20	2	4	2	14.0000	0.3	R (11)
6	7	112	66	4	6	30	58.3333	0.11616	R (13)
7	11	132	40	3	5	6	30.8	0.23	R (29), MH (5)
8	11	176	66	4	6	18	55.0000	0.16667	R (30)
9	13	104	20	2	4	1	13.0000	0.35	R (37), MH (3)
10	16	256	66	4	6	12	53.3333	0.19192	R (47), MH (10)
11	21	252	40	3	5	3	29.4	0.265	R (58), MH (7)
12	31	496	66	4	6	6	51.6667	0.21717	R (75), MH (12)

**The symbols R( ) and MH( ) denote the reference number in Raghavrao [30] and Marshal Halls [35] list.

ACKNOWLEDGMENTS

We are grateful to the anonymous referees for their constructive comments and valuable suggestions.

REFERENCES

1. Agrawal, H.L. and Prasad, J. (1982) Some Methods of Construction of Balanced Incomplete Block Designs with Nested Rows and Columns. Biometrika, 69, 481-483. http://dx.doi.org/10.1093/biomet/69.2.481

2. Caliski, T. (1971) On Some Desirable Patterns in Block Designs. Biometrika, 27, 275-292. http://dx.doi.org/10.2307/2528995

3. Hanani, H. (1975) Balanced Incomplete Block Designs and Related Designs. Discrete Mathematics, 11, 255-269. http://dx.doi.org/10.1016/0012-365X(75)90040-0

4. Shrikhande, S.S. and Raghavarao, D. (1963) A Method of Construction of Incomplete Block Designs. Sankhya, A25, 399-402.

5. Jones, R.M. (1959) On a Property of Incomplete Blocks. Journal of the Royal Statistical Society, B21, 172-179.

6. Puri, P.D. and Nigam, A.K. (1975) On Patterns of Efficiency Balanced Designs. Journal of the Royal Statistical Society, B37, 457-458.

7. Rao, V.R. (1958) A Note on Balanced Designs. Annals of the Institute of Statistical Mathematics, 29, 290-294. http://dx.doi.org/10.1214/aoms/1177706729

8. Williams, E.R. (1975) Efficiency Balanced Designs. Biometrika, 62, 686-689. http://dx.doi.org/10.2307/2335531 R. Awad, S. Banerjee 1170

9. Van Lint, J.H. (1973) Block Designs with Repeated Blocks and (b; r; λ) = 1. Journal of Combinatorics Theory, A15, 88-309.

10. Ceranka, B. and Graczyk, M. (2007) Variance Balanced Block Designs with Repeated Blocks. Applied Mathematical Sciences, 1, 2727-2734.

11. Ceranka, B. and Graczyk, M. (2009) Some Notes about Efficiency Balanced Block Designs with Repeated Blocks. Metodoloski Zvezki, 6, 69-76.

12. Ghosh, D.K. and Shrivastava, S.B. (2001) A Class of Balanced Incomplete Block Designs with Repeated Blocks. Journal of Applied Statistics, 28, 821-833. http://dx.doi.org/10.1080/02664760120074915

13.   Hedayat, A. and Federer, W.T. (1972) Pairwise and Variance Balanced Incomplete Block Designs. Annals of the Institute of Statistical Mathematics, 26, 331-338. http://dx.doi.org/10.1007/BF02479828

14.   Hotelling, H. (1944) Some Improvements in Weighing and Other Experimental Techniques. Annals of Mathematical Statistics, 15, 297-306. http://dx.doi.org/10.1214/aoms/1177731236

15.   Yates, F. (1935) Complex Experiments. Supplement to the Journal of the Royal Statistical Society, 2, 181-247.

16.   Banerjee, K.S. (1948) Weighing Designs and Balanced Incomplete Blocks. Annals of Mathematical Statistics, 19, 394-399. http://dx.doi.org/10.1214/aoms/1177730204

17.   Banerjee, K.S. (1975) Weighing Designs for Chemistry, Medicine, Economics, Operations Research, Statistics. Marcel Dekker Inc., New York.

18.   Dey, A. (1969) A Note on Weighing Designs. Annals of the Institute of Statistical Mathematics, 21, 343-346. http://dx.doi.org/10.1007/BF02532262

19.   Dey, A. (1971) On Some Chemical Balance Weighing Designs. Australian Journal of Statistics, 13, 137-141. http://dx.doi.org/10.1111/j.1467-842X.1971.tb01252.x

20.   Raghavarao, D. (1959) Some Optimum Weighing Designs. Annals of Mathematical Statistics, 30, 295-303. http://dx.doi.org/10.1214/aoms/1177706253

21.   Ceranka, B. and Graczyk, M. (2001) Optimum Chemical Balance Weighing Designs under the Restriction on the Number in Which Each Object Is Weighed. Discussiones Mathematicae: Probability and Statistics, 21, 113-120.

22.   Ceranka, B. and Graczyk, M. (2002) Optimum Chemical Balance Weighing Designs Based on Balanced Incomplete Block Designs and Balanced Bipartite Block Designs. Mathematica, 11, 19-27.

23.   Ceranka, B. and Graczyk, M. (2004) Ternary Balanced Block Designs Leading to Chemical Balance Weighing Designs for v + 1 Objects. Biometrica, 34, 49-62.

24.   Ceranka, B. and Graczyk, M. (2010) Some Construction of Optimum Chemical Balance Weighing Designs. Acta Universitatis

Lodziensis, Folia Economic, 235, 235-239.

25. Awad, R. and Banerjee, S. (2013) Some Construction Methods of Optimum Chemical Balance Weighing Designs I. Journal of Emerging Trends in Engineering and Applied Sciences (JETEAS), 4, 778-783.

26. Awad, R. and Banerjee, S. (2014) Some Construction Methods of Optimum Chemical Balance Weighing Designs II. Journal of Emerging Trends in Engineering and Applied Sciences (JETEAS), 5, 39-44.

27. Caliski, T. (1977) On the Notation of Balance Block Designs. In: Recent Developments in Statistics, North-Holland Publishing Company, Amsterdam, 365-374.

28. Kageyama, S. (1974) On Properties of Efficiency Balanced Designs. Communications in Statistics-Theory and Methods, 9, 597-616.

29. Banerjee, S. (1985) Some Combinatorial Problems in Incomplete Block Designs. Unpublished Ph.D. Thesis, Devi Ahilya University, Indore.

30. Raghavarao, D. (1971) Constructions and Combinatorial Problems in Designs of Experiments. John Wiley, New York.

31. Shah, K.R. and Sinha, B.K. (1989) Theory of Optimal Designs. Springer-Verlag, Berlin, Heidelberg. http://dx.doi.org/10.1007/978-1-4612-3662-7

32. Wong, C.S. and Masaro, J.C. (1984) A-Optimal Design Matrices $X = \left(x_{ij}\right)_{N \times n}$ with $x_{ij} = -1,0,1$. Linear and Multilinear Algebra, 15, 23-46. http://dx.doi.org/10.1080/03081088408817576

33. Jacroux, M., Wong, C.S. and Masaro, J.C. (1983) On the Optimality of Chemical Balance Weighing Designs. Journal of Statistical Planning and Inference, 8, 231-240. http://dx.doi.org/10.1016/0378-3758(83)90041-1

34. Ceranka, B. and Graczyk, M. (2007) A-Optimal Chemical Balance Weighing Design under Certain Conditions. Metodoloski Zvezki, 4, 1-7.

35. Hall Jr., M. (1986) Combinatorial Theory. John Wiley, New York.

Design of Local Roadway Infrastructure to Service Sustainable Energy Facilities

Karim A AbdelWarith[1], Panagiotis Ch Anastasopoulos[2], Wayne Richardson[3], Jon D Fricker[1],
and John E Haddock[1]

[1]School of Civil Engineering, Purdue University: Indiana Local Technical Assistance Program, 550 Stadium Mall Drive, West Lafayette, IN 47907, USA

[2]Department of Civil, Structural and Environmental Engineering, Institute for Sustainable Transportation and Logistics, University at Buffalo, The State University of New York, 241 Ketter Hall, Buffalo, NY 14260, USA

[3]Bertsch-Frank & Associates, LLC, 4630 W. Jefferson Blvd. #6, Fort Wayne, Indiana 46804, USA

ABSTRACT

Background

This paper aims to identify specific local roadway infrastructure design guidelines associated with the construction and operation of sustainable energy source facilities, such as ethanol plants, biomass plants, and wind farm facilities.

Methods

Data associated with sustainable energy facility traffic in Indiana were collected to develop Excel-based tools (worksheets) and assist local agencies in the design of pavements in the proximity of ethanol plants, biomass plants, and wind farms.

Results

To that end, a simple procedure is presented, which provides a design capable of withstanding heavy traffic loads, while, at the same time, quantifies the effects that new sustainable energy source facilities may have on local road networks. The procedure is accompanied by two MS Excel-based software tools that can be used in the design of local roads adjacent to such sustainable energy facilities.

Conclusion

The developed worksheets can serve as a hands-on tool to assist local government engineers in evaluating and in quantifying the probable effects of the construction and operation of a sustainable energy facility in their jurisdiction.

BACKGROUND

Renewable, sustainable energy sources are being developed at a record pace throughout the USA and globally, with multidimensional benefits, as they have the potential to boost local economies and generate new jobs [1-20]. In Indiana, energy corporations have invested in three main types of sustainable energy sources, namely, ethanol, wind, and biomass energy, and have built numerous wind farms and ethanol and biomass plants. It is expected that the number of plants and wind farms will triple by 2022 [21]. Increased loads, increased traffic, or both can negatively affect road networks (with respect to the existing infrastructures, the environment, the aesthetics of the local communities, and the safety of the neighboring residents) when sustainable energy projects are introduced into a community [22-28]. Wind farm construction increases the loads on roads leading to and from the wind farm during turbine construction, but once the turbines have been constructed, there is nearly no increase in traffic [26,27]. Conversely, when a fixed-point energy source that must be serviced by trucks is constructed, such as an ethanol or biomass plant, it results in additional traffic, and on many occasions, increased loads [22-25]. While it may be possible to mitigate these effects by the use of barge or rail [29], at some point, the road network will need to be used to move the turbine components, or the biomass or ethanol products.

In Indiana, ethanol plants, biomass power plants, and wind farms are typically built in rural areas. Most local road networks were not designed or constructed to accommodate the increased traffic and loads produced by such facilities. When sustainable energy developers decide to locate facilities within a given governmental entity, local officials need to have a sound understanding of the proposed facilities' probable effects on their local road network and some methods to quantify those effects. The local highway engineers and supervisors also need to be familiar with the resulting traffic and load problems associated with these facilities and be in a position to make decisions as to which pavement structure is needed to bear such heavy loads and traffic near the facilities.

Previous research in biomass and ethanol usage has pointed out the importance of designing access roads or considering the capacity of access roads to the plant [30, 31]. However, the existing literature, to the authors' knowledge, does not illustrate how loads can be calculated or access roads be designed for these facilities. Furthermore, research that has focused on wind farm technology suggests that local roads should handle the heavy construction loads from the wind mill parts [32-36]. On one hand, the focus has been on detailed design methodologies aiming to handle these loads, while on the other hand, guidelines to develop temporary access roads for wind farms were also presented [37].

As illustrated herein, the aforementioned design problems are solved using existing design guides, such as the American Association of State Highway and Transportation Officials (AASHTO) method. However, such methods may often be intricate, which would inevitably require consultation with expert designers. Even though third-party expert consultation is welcome, local authorities generally do not have the necessary funds for this process.

This paper aims to develop tools that can be used by local government agencies in quantifying the effects of proposed sustainable energy projects on their local road networks. The tools are designed and developed, bearing in mind that local agencies do not typically employ personnel with specific expertise in pavement analysis and design. These tools are therefore expected to assist local agency personnel in determining appropriate pavement sections and quantifying their costs. The paper is organized as follows. First, background information on renewable energy resources are given, along with biofuel transportation practices. Next, the method and data are presented, followed by the design development description of the proposed tool. Finally, the tool validation results are discussed.

The contribution of this paper lies in the development of local roadway infrastructure design guidelines associated with the construction and operation of sustainable energy source facilities, such as ethanol plants, biomass plants, and wind farm facilities. The proposed procedure is designed to be simple and is accompanied by hands-on tools to assist local government engineers in evaluating and in quantifying the probable effects of the construction and operation of a sustainable energy facility in their jurisdiction. Therefore, the procedure is anticipated to provide designs capable of withstanding heavy traffic loads, while, at the same time, it has the potential to quantify the effects that new sustainable energy source facilities may have on local road networks.

Renewable Energy Resources

In order to better comprehend the local effects of the construction and operation of sustainable energy projects, such as ethanol, biodiesel, biomass, and wind energy, some background information is briefly presented. Ethanol can be produced from a number of agricultural products, such as sugar and starch [38]. The ethanol production process yields several byproducts, such as dried distillers grains with solubles   (DDGS), which are a high-nutrient feed valued by the livestock industry [39]. Ethanol demand is difficult to capture, given its dual nature, i.e., being both an additive to and a substitute for gasoline. However, the market for ethanol significantly increased (over 500%) when flexible fuel vehicles (FFV) were made available to the public [21, 39, 40]. In Indiana, there are 11 ethanol plants, plus 2 under construction (see Figure 1).

Figure 1: Ethanol plants in Indiana.

Biomass is a plant matter grown to generate electricity or produce heat, with agricultural waste being the most common type of solid biomass that can be used as a source of energy [41]. Biomass currently provides about 10% of the world's primary energy supplies, most being used in developing countries in the form of fuel wood or charcoal for heating and cooking [42,43]. In the USA, 85% of the wood production industry waste is used for power generation, with approximately 80 operating biomass power plants (and 40 operable but idle plants) located in 19 states across the country [44]. Demand for power derived from biomass is generally increasing, having surpassed hydropower as the largest domestic source of renewable energy [45]. In Indiana, there is currently one biomass plant near Milltown in Crawford County [46].

The power of the wind can be harnessed and converted to electricity by the use of tower-mounted wind turbines. Wind turbines can be used to produce electricity for a single home or building, or they can be connected to an electricity grid for more widespread electricity distribution. Wind energy is not only 'green' but also cost effective when compared to other sources of electricity in the USA. The growing wind power market has attracted many energy corporations to the field [47, 48]. In the USA, not all regions have wind speeds that are high enough to support wind energy production [49]. However, a recent study showed that building wind farms on only 3% of the area of the USA will produce enough electricity to meet all US energy demands [50]. In Indiana, there are currently 18 wind farms in operation, with over a 1,500 MW of wind electricity-generating capacity [51]. Indiana has the potential to produce 150,000 MW of electricity from wind farms [51].

Transportation of Biofuels

Biofuels can be transported by trucks, rail, or barge. Trucks are used when the material needs to be transported from one mode to another. Rail transportation is effective for long hauls, while barges are the least expensive transport method. Barges can carry large amounts to export terminals, and then ocean vessels are used to carry them to foreign markets. Transportation of biofuels using pipelines is limited in the USA due to the adverse impact of the former (mostly their chemical properties) on the pipeline integrity and safety. Also, pipelines are not largely available where biofuel plants are located. However, pipelines are a feasible option for the transportation of conventional fuel types. Air transport is not a viable option either, due to its high cost.

From a capacity standpoint, a truck can accommodate approximately 25,000 liters, a railcar approximately 95,000 liters, and a barge approximately 1,500,000 liters. On the other hand, it would not be economically effective to use rail or barges to transport biofuels to short distances, such as locations within less than 80-km distance from the facility [52]. In general, truck transportation is considered to be an efficient mode of transportation up to a distance of 500 km [52]. A railcar can transport the freight 2.5 times farther than a truck, for the same cost per liter, whereas barges can move freight across long distances and oversees (e.g., from the Midwest to the Gulf).

Rail is used to transport 41% of US corn exports and 14% of corn domestically [53]. In 2005, rail was the primary transportation mode for ethanol, shipping 60% of ethanol produced, or approximately 11 billion liters. In comparison, trucks shipped 30% and barges 10%. Although trucks are used to ship most of the corn used by ethanol plants, some of the newer and larger plants use rail for inbound corn shipments [54].

Barges move approximately 5% to 10% of ethanol, in addition to the DDGS and fertilizers necessary for the production of corn. Barges also move 44% of all grain exports. In 2007, barges moved 55% of corn to ports, and 1% of corn to processors, feed lots, and dairies [54]. An issue with barge transportation is related to the occasional inadequacy of water depths that can lead to higher transportation costs. Seasonal effects on barge transportation may also decrease the barge's moving capacity (e.g., at a 2.75-m draft, if a barge has 1,500 tons of capacity, every 2.5 cm of reduced draft will result in 17 tons of reduced capacity) [53].

In the Midwest, inbound corn being delivered to the processing facility is most typically delivered by trucks from corn farms within an 80-km radius. Standard gasoline tanker trucks (DOT MC 3066 Bulk Fuel Haulers) are typically used to ship ethanol outbound from the plants to the blending terminals. The total number of independently operated tank trucks is approximately 10,000, excluding the tanker truck fleets that are owned by petroleum companies [55].

METHODS

The objective of this paper is to develop a design methodology to assist local agencies in designing suitable pavements for sustainable energy projects served by local roads. In order to ensure a reliable pavement design, the first step is to collect accurate data with respect to the operation and traffic generation of sustainable energy projects. Development of the design tools is completed based on the following criteria: the tools should (a) be simple and easy to use; (b) require minimum input from the user and, at the same time, allow for more experienced users to input more detailed data; and (c) be able to produce several alternative pavement sections, when applicable. The output of the study involves worksheet-based pavement design procedures, one for ethanol and biomass plants, and a second for wind farms. These tools offer a user-friendly interface and several levels of input regardless of the expertise of the user.

The design development phase is based on various design guides and design elements that have been proven useful in the design of specialized pavements for sustainable energy projects. The design guides considered are the AASHTO Pavement Design Guide (for flexible and rigid pavements and for low-volume road design), the Asphalt Institute Pavement Design Guide, the Mechanistic-Empirical Pavement Design Guide (MEPDG), and the Portland Cement Association (PCA) [56-59]. Reviewing of these sources shows that rigid pavement design is not typically used in the design of local low-volume roads. Thus, the AASHTO Rigid Pavement Design Guide and the PCA Pavement Design were not utilized in the proposed overall design methodology. The MEPDG was found to be complex and was not geared toward low-volume roads and was therefore not used either. The AASHTO flexible pavement design was utilized due to its simplicity, versatility, and robustness. The AASHTO low-volume road design was also utilized in the ethanol and biomass worksheet. As for the wind farm worksheet, the Asphalt Institute's Manual Series No. 23 (MS-23), 'Thickness Design: Asphalt Pavements for Heavy Wheel Loads' was the only design guide that addressed the large, one-time loads expected during the construction of wind farm facilities [60].

As a final step, the proposed design methodology is tested to ensure that it produces realistic results. Ideally, the proposed design methodology would be validated by building a road conforming to the design methodology, then monitor it over several years, and determine whether it fails prematurely. Obviously, this falls out of the scope of the current study. Instead, the proposed methodology is validated by comparing it to in-service designs currently servicing sustainable energy projects. If the simplified proposed design provides an output that falls close to the outputs of the designs in place, the proposed design is considered adequate. This, of course, does not guarantee an optimum design; it suggests, though, that the developed designs are approximating actual design results.

Interviews with Local Officials

As part of the data collection process, interviews were conducted with Indiana's local road agency representatives in counties where biomass plants, ethanol plants, or wind farm facilities are located. The interviews entailed a set of questions about the provided provisions in anticipation of the increased traffic and the current condition of the road network. To that end, 12 counties that have ethanol and biomass plants were interviewed. Of those 12, only 4 had performed any type of upgrade to their local roads in anticipation of increased traffic. Table 1 summarizes the representatives' responses.

Table 1: Summary of local agency survey responses

County	Is the plant operational?	General response	Upgrades performed
Cass	Yes	No response	Unsure
Grant	No	No upgrades were performed on county roads. The nearest state highway was widened to accommodate the large-radius turning paths of long trucks. The plant does not have a county access road.	No
Henry	Yes	The plant is located right adjacent to a state highway; thus, there was no need for any upgrades. However, roads are deteriorating quickly, and there is no funding from the state or other sources.	No
Jasper	Yes	No provisions were needed because the plant is located adjacent to a state highway.	No

Jay	Yes	Upgrades were performed on county roads. The plant created a tax increment financing (TIF) district, and the new roads were paid for using the money from the bonds sold. Upgrades included widening and resurfacing of a section of a county road. The main problem is that truck drivers do not always use that route; thus, other roadways may deteriorate.	Yes
Kosciusko	Yes	The Highway supervisor expressed concern about the highways. Attempts were made to get funds to perform repairs. No legal agreement between the plant and the county was made	No
Lake	No	No response	No
LaPorte	No	No response	No
Madison	Yes	No response	Unsure
Montgomery	Yes	No response	No
Posey	Yes	There are two plants; one is adjacent to a state highway, the other is not. The latter required road upgrades. The upgrades were paid for through setting up a TIF district. Also, there were two low-volume roads that the plant wanted to build a bridge over. The county engineers were able to reach to an agreement with the commissioners to close these two roads, saving the expense of building an overpass. In return, the county received one million dollars which they used to repair and upgrade highways. The upgrades included mainly 5 to 7.5 cm of resurfacing on access roads.	Yes
Putnam	No	No response	No
Randolph	Yes	The plant built a private access road to a county road that was partly upgraded. The county is currently working on an agreement with the plant to upgrade the roads used by farmers.	Yes
Shelby	Yes	No response	Unsure
Wabash	Yes	The county established a TIF district in the area to be developed. The county performed road upgrades which included digging up the existing pavement, placing a 33- to 38-cm Portland cement-stabilized soil and HMA on top. The project cost was $1.2 million. The county was later reimbursed by the plant (as agreed before the start of the project by selling TIF district bonds). The county also received an economic development stimulus from the state of Indiana.	Yes

| Wells | Yes | There were no upgrades performed. However, there were discussions at the time of construction that the plant had a budget set for upgrading the roadway. Due to technical difficulties on the county/city side, the roads were not upgraded. The plant did not spend any of the allocated budgets. The county engineers tried to mitigate the damage by channelizing the truck traffic produced by the plant onto roadways that could accommodate the traffic. The county engineers provided this channelization through verbal coordination with truck companies and drivers. The highway supervisor stated that the truck companies were very cooperative. | No |

AbdelWarith et al.

AbdelWarith et al. Energy, Sustainability and Society 2014 4:14

Traffic associated with ethanol and biomass plants can be classified as follows: (a) incoming traffic handling raw materials and (b) outgoing traffic handling product distribution. Incoming traffic is mainly composed of trucks, while outgoing traffic is composed of rail and truck traffic, in most cases. For this reason, plants are typically located near major highways and rail sites. In Indiana, all plants are located within 4 km from the nearest state highway or interstate and within 1.1 km from the nearest rail freight facility. Of all the plants, 85% are located within 1.6 km of a state highway. Of all operating ethanol and biomass plants in Indiana, 30% are located adjacent to a main highway; whereas 23% of all operating ethanol and biomass plants in Indiana have rail tracks leading into their facility. On average, the plants in Indiana are 0.87 km away from a state highway and 0.5 km from a railroad.

County officials and plant managers expected that all truck traffic would use the nearby state highways or interstates. Thus, no significant upgrades were performed on local roads. In many cases, the expectation that trucks would utilize the state highways or interstates was not validated. Truck drivers use the shortest route unless otherwise instructed, which may or may not be a state highway or interstate. Also, farmers delivering raw materials to the plant came from all directions. This entailed utilizing county roads.

Local county highway representatives were interviewed in several counties in Indiana, in which wind farms are located. In both cases, the wind farm developers signed a road use agreement with the county specifying that the developers are responsible for the road condition. The developers agreed to return the roadways used in the wind farm construction process to their original condition and further performed significant upgrades to the local roads. However, detailed information was only available from White County, which was used in the validation process.

Note that road use agreements typically include warranty clauses, which provide an assurance to the owner that the product/service will serve its useful life without failure, and if it does not, the contractor will repair or replace the product (for specifics on roadway preservation through public-private partnerships, see [61-69]). In the case of White County, a 2-year warranty was defined. Benton County defined a 1-year warranty on roads and a 5-year warranty on drainage.

Data

Biomass and Ethanol Plants

The amount of traffic associated with an ethanol plant is directly related to the plant's capacity, most often measured in millions of liters per year (MLY). Because the plants are normally located to take advantage of locally produced raw materials, in this case corn, nearly all of the incoming raw material is delivered to the plant by tractor-trailers. The outgoing products are ethanol and DDGS. In Indiana, nearly all of the ethanol leaves the plant by train. The DDGS may be transported by train or truck, depending on local livestock markets. Plant capacities, amount of raw materials consumed, and plant production rates for each of the ethanol plants in Indiana are summarized in Table 2.

Table 2: Indiana ethanol plant data

Plant	County	Annual liters of ethanol produced (millions)	Annual bushels of corn used (millions[a])	Corn used per liter of ethanol produced (bushels)	Annual tons of DDGS produced (thousands)	Annual tons of DDGS produced per liters of ethanol produced (millions)
Anderson Ethanol	Cass	416	39	0.0937	354	0.850
Cardinal Ethanol	Randolph	379	37	0.0977	321	0.848
Central Indiana Ethanol	Grant	151	15	0.0991	145	0.958
Indiana Bio-Energy	Wells	416	37	0.0889	321	0.771
Iroquis Bio-Energy Company	Jasper	151	15	0.0991	129	0.852
New Energy Corp.	St. Joseph	379	37	0.0977	328	0.866
POET	Jay	246	24	0.0975	193	0.784
POET	Madison	227	22	0.0969	193	0.850
POET	Wabash	246	24	0.0975	209	0.849
Valero Energy (formerly Vera-Sun)	Montgomery	379	37	0.0977	350	0.925
Altra (not operating)	Putnam	227	22	0.0969	192	0.845
Abengoa Bioenergy	Posey	333	32	0.0961	282	0.847
Total		3,551	341	1.16	3,017	10.25
Average		296	28	0.10	251	0.85

[a]One bushel of corn weighs about 25 kg.

AbdelWarith et al.

AbdelWarith et al. Energy, Sustainability and Society 2014 4:14

The amount of raw material consumed by a biomass plant is governed by the plant's capacity, the amount of electricity it can produce, and the plant's efficiency. Capacity is measured in megawatt electrical (MWe), while efficiency by the heat production rate is measured in watts per kilowatt-hour (W/KWh). Each material, when burned, produces a specific amount of heat energy measured in watts per kilogram (W/kg). Herein, a constant value of 2,746 W/kg for all agricultural byproducts is adopted from Wilt see [70]. The average heat rate of 140 biomass plants listed in the National Electric Energy System Database [71] was also used, which was calculated to be 4,462 W/KWh.

Unlike ethanol plants, biomass plants do not produce loaded, outgoing traffic. Raw materials are shipped to the plant and burned to generate electricity. The type of input materials varies and can be divided into four main types: woody plants, herbaceous plants/grasses, aquatic plants, and manures [72]. According to the local Indiana farmers, woody and herbaceous plants are the most commonly used raw materials in Indiana biomass plants, with the most typical being corn stover, wood chips, sawdust, and baled straw [72-74]. Each material has a different density, as shown in Table 3. The less dense the material, the more space per kilogram it occupies; thus, more trucks are needed to transport less dense materials. This was taken into consideration when calculating loads associated with biomass plant operation.

Table 3: Biomass raw material densities

Material	Density (kg/m³)
Corn stover	128.15
Wood chip	200.23
Sawdust	120.14
Baled straw	150.57

AbdelWarith et al.

AbdelWarith et al. Energy, Sustainability and Society 2014 4:14

Wind Farms

The increased truck traffic associated with wind farm facilities is mostly limited to construction traffic, which can be divided into transportation of construction materials (concrete, aggregates, and steel reinforcing), transportation of construction equipment (cranes), and transportation of wind turbine components (nacelle, rotor, blades, and tower sections) The construction materials represent the heaviest loads per truck axle. The turbine components can be heavy, but additional axles are added to the truck trailer as needed so as to comply with axle weight limits. In most cases, the length of the turbine components is the most critical concern. Wind turbine components, such as blades and tower sections, are extremely long and require long trucks to haul them. Blades are typically 45 m in length and weigh 11,340 kg [26]. While the weight is distributed over a large number of axles, the challenge is making sure that trucks have sufficient turning radii when using local roads. Table 4 summarizes the weight and truck axles needed for various wind turbine components [71].

Table 4: Truck information for various wind turbine components

Component	Weight (kg)	Longest dimension (m)	Minimum number of truck axles needed to carry component	Weight per axle (kg/axle)	Weight per tire (kg/tire)
Base section	41,958	14.66	4	10,490	2,622
Lower-middle section	41,241	19.81	6	6,874	3,789
Upper-middle section	28,111	19.90	6	4,685	2,583
Top section	28,876	22.59	6	4,813	2,653
Hub	17,010	3.84	3	5,670	3,125
Blades	6,486	33.99	6	1,081	596
Rotor	32,024	70.47	6	5,337	2,942
Nacelle	57,153	8.81	3	19,051	10,500

AbdelWarith et al.

AbdelWarith et al. Energy, Sustainability and Society 2014 4:14

The erection of wind turbines includes two major activities: off-loading and stacking out. Off-loading normally requires a 200-ton crawler or hydraulic crane. Stack-out requires a 400-ton crane [75]. Both cranes are transported in pieces and assembled on site. Table 5 summarizes the weight of each component and the number of truck axles required to carry it. Each crane is assembled in 20 to 25 truck trips, which are performed at least twice (assembling and disassembling) in the project lifetime, regardless of the number of wind turbines being built [26,76,77].

Table 5: Crane components

Equipment	200-ton crane			400-ton crane		
	Weight (kg)	Number of axles	Weight per tire (kg/tire)	Weight (kg)	Number of axles	Weight per tire (kg/tire)
Basic crane	39,689	4	2,480.58	39,612	4	2,475.76
Car-body and adapter	N/A	N/A	N/A	28,161	3	2,346.77
Crawler assembly	19,622	4	1,226.40	32,665	4	2,041.59
Counterweight tray	9,548	3	795.68	19,958	3	1,663.17
Upper-center counterweight	10,659	3	888.28	8,165	3	680.39
Upper-side counterweight	7,938	3	661.49	6,804	3	566.99
Lower car-body counterweight	9,979	3	831.59	13,608	3	1,133.98
Upper car-body counterweight	8,165	3	680.39	N/A	N/A	N/A
9 m boom butt	4,910	3	409.18	21,609	3	1,800.76
12 m boom top	2,544	3	212.02	5,595	3	466.25
3 m boom insert	971	3	80.89	N/A	N/A	N/A
6 m boom insert	1,397	3	116.42	2,563	3	213.57

AbdelWarith et al.

AbdelWarith et al. Energy, Sustainability and Society 2014 4:14

As mentioned earlier, the heaviest load associated with wind farm construction is the construction materials. Wind tower foundations require 282 to 480 yd³ of concrete and 20 to 38 tons of steel reinforcement [75]. Truck traffic is also generated by the need to transport aggregates to the site. Table 6 presents the number of trucks needed to construct the foundation of a single turbine and the weights of each construction material used [23]. Finally, data on local pavement construction materials were collected from local suppliers and used in the proposed design process. One of the design outputs is the cost of the recommended pavement. The specific gravity and cost data collected and utilized in the analyses are illustrated in Table 7.

Table 6: Wind-tower foundation construction materials

Construction material	No. of trucks required	Truck loads (kg)	Load per tire (kg)
Aggregate	10	22,680	5,670
Concrete	20 to 40	22,680	5,670
Steel	1	18,144	4,536

AbdelWarith et al.

AbdelWarith et al. Energy, Sustainability and Society 2014 4:14

Table 7: Pavement construction materials, specific gravities, and costs

Material	Specific gravity	Density (kg/m³)	Tons/lane-km/cm	Price/ton/lane	Price/lane-km/cm
Hot mix asphalt	2.65	2,643	106.66	$90	$9,599
Compacted dense aggregate	2.75	2,739	110.82	$13	$1,441
Coarse aggregate	2.45	2,435	98.83	$9	$889
Excess excavation					$889

AbdelWarith et al.

AbdelWarith et al. Energy, Sustainability and Society 2014 4:14

RESULTS AND DISCUSSION

Design Development

Two MS Excel-based pavement design procedures (Excel worksheets) were developed, one for ethanol and biomass plants, and a second for wind farms. Both worksheets follow the logic outlined in Figure 2.

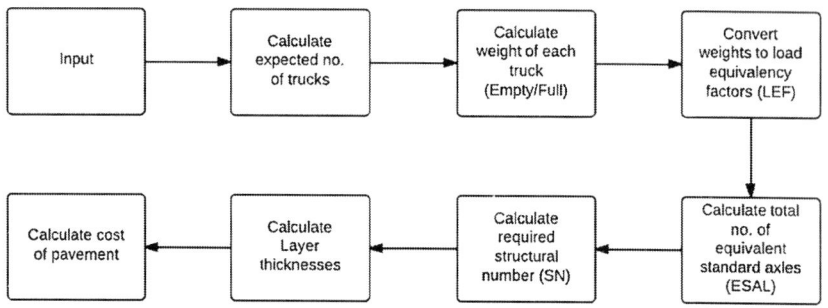

Figure 2: Conceptual illustration of the pavement design procedures.

Ethanol and Biomass Design Specifications

The ethanol and biomass design procedure is based on the standard AASHTO Pavement Design Guide [56]. As discussed previously, ethanol and biomass plants produce significant traffic, which is mainly composed of trucks. The traffic produced is typically above one million equivalent single axle loads (ESAL) over a 20-year period, which merits a design governed by pavement fatigue standards.

There are primary and secondary design inputs, which are summarized in Table 8. Primary inputs are project-specific, which vary significantly from one project to another. Secondary inputs are approximately constant across projects, either due to the nature of these factors, or because they tend to be standardized. These inputs are set at default, but can be customized by the user if desired.

Table 8: Primary and secondary inputs needed for the ethanol and biomass procedure

Primary inputs	Description	Secondary inputs	Description
Plant capacity	This value represents the maximum amount of biofuel that could be produced by the plant considered. This value should be reported in million liters per year for ethanol plants and megawatts electrical (MWe) for wind farms.	Yearly growth factor	If the plant is expected to increase its capacity in the future, the designer may add a reasonable growth factor. However, it is uncommon for a plant to be built and have its capacity increased later on in its service life. The default value is 0%.
Capacity factor for biomass plants	Biomass plants produce power but are not exclusively consistent in doing so. The capacity factor is a percentage that represents the average power output of a plant. It ranges from 15% to 100%. When this value is unknown, an average value of 67% is assumed.	Ethanol plant products and raw materials	Amount of corn hauled to the plant by trucks, as opposed to be transported by another means, or grown within the plants' grounds, thus not using any local roads. Therefore, the default value assigned is 100%.
			Ethanol hauled from the plant by trucks, as opposed to stored or sold locally. Most ethanol plants are located close to rail freight tracks. For this reason, most (if not all) of the ethanol production will be shipped by rail. The assigned default value is conservatively set to 20%.
			DDGS hauled from the plant by trucks. This has a default value of 20%.

Design period	The amount of time the road is expected to remain in service without major rehabilitation. This value is typically 20 years; however, for county roads, it can be lower.	Biomass fuel type	Biomass fuel can be produced using various components and ratios of these components. Different component raw materials have different weights. It is important to determine the different percentages of each raw material to avoid underestimating the weight of incoming trucks. The four typical components are corn stover, wood chips, saw dust, and baled straw. Each is set at a default value of 25%.
California bearing ratio (CBR)	This value reflects the strength of the underlying soil. To get the actual field CBR of the soil, soil bores need to be drilled in the construction location. However, the value can be closely estimated by knowing the type of soil in the area. Highway supervisors can resort to previous experience or soil maps to determine the soil type in the area.	Reliability (R %)	The designer should choose the level of reliability of the design. For local county roads, the reliability is typically low. This value ranges from 50% to 99%; 75% is the default value.
		Terminal serviceability index (Pt)	This is the value that reflects the condition of the pavement at the end of its service life. This value ranges from 3 (for major highways) to 1.5 (minimum). The default value is 2, as recommended for county roads by the AASHTO design guide [49].
		Overall standard deviation	This number reflects the variability within the pavements' materials. It typically ranges from 0.3 to 0.5. A value of 0.5 is recommended by the AASHTO design guide [49] and is set as the default value.

AbdelWarith et al.

AbdelWarith et al. Energy, Sustainability and Society 2014 4:14

In order to simplify the design process, a number of assumptions are made. It is important to note that these assumptions could be easily changed by the user if so desired. The following are the assumptions considered in this design procedure:

- The capacity of a typical truck is assumed to be 90.6 m³.
- An average number of bushels of corn (0.091 bushels/liter of ethanol) is used to calculate the amount of corn needed to supply the ethanol plant:
- Unit analysis (1 bushel = 25 kg)
- For one million liters of ethanol, 91,000 bushels are needed (0.091 × 1,000,000); using a 25-ton (25,000 kg) truck, the number of trucks needed per day in a year is equal to 91,000 × 25 / (25,000 × 365) = 0.25.
- Using unit analysis, it is found that for each one million liters per year of ethanol to be produced, 0.25 trucks per day are needed to supply the plant. The number of ethanol tanker trucks is calculated, assuming that tankers haul on average 30,000 liters. Using an ethanol density of 0.789 kg/liters, the weight per truck is calculated as 23,000 kg [55].
- One million liters of ethanol has a byproduct of 845 tons of DDGS and a truck can haul approximately 25 tons of distiller's grain [55], leading to a factor of 0.0926 trucks per day:
- Unit analysis
- For one million liters of ethanol, 845 tons of DDGs is needed; using a 25-ton (25,000 kg) truck, the number of trucks needed per day in a year is equal to: 845 / (25 × 365) = 0.0926.
- The ESAL of the trucks are calculated using the fourth power law load equivalency factor (LEF). This law uses the weight over a constant based on axle type raised to the fourth power [78]:

$$LEF = \left(\frac{\text{Weight of axle}}{\text{Weight of constant}} \right)^4$$

$$\tag{1}$$

$$ESAL = LEF \times \text{Number of vehicles with that axle weight}$$

$$\tag{2}$$

For a single axle, the constant used is 8,160 kg. For a tandem axle, the constant used is 15,060 kg.where W_{18} = ESAL (reflects road traffic),

Z_R = standard normal deviation (reflects the reliability of design), S_0 = standard deviation (reflects variability of pavement material), ΔPSI = reflects the difference between pavement condition right after construction and the end of its service life, and M_R = resilient modulus (PSI) (reflects subgrade strength).

- For the biomass facility, the weight of product produced is calculated using typical heat and production rates: 4,103 W/kWh heat rate and 2,746 W/kg fuel production rate.

- For biomass facilities, the densities of the raw materials are used to calculate the weight and number of trucks going into the plant. The densities of these raw materials are listed in Table 3.

- The AASHTO design guide's [56] ESAL equation is used to find the pavement's structural number:

$$\log W_{18} = Z_R + S_0 + 9.36\,(\log(\text{SN}+1)-0.2)$$
$$+ \frac{\log \frac{\Delta\text{PSI}}{4.2-1.5}}{0.4 + \frac{1.094}{(\text{SN}+1)^{5.19}}} + 2.32(\log(M_R)-8.07).$$

(3)

The design normally consists of three layers. Layer coefficients, a, are assumed to take the following values: $a_1 = 0.4$ (in the range of 0.2 to 0.4; 0.4 is typically used), $a_2 = 0.14$ (recommended by AASHTO for granular base layers), and $a_3 = 0.11$ (recommended by AASHTO for granular sub-base layers). Drainage coefficients, m, are assumed to take the following values: $m_2 = 0.8$ and $m_3 = 0.8$ (these values reflect fair drainage with 75% reliability). Initial pavement serviceability (Po) is assumed to be 4.2 (this value is the AASHTO design guide [56] recommended value for flexible pavements). The resilient modulus introduced in the design's internal calculations is a function of the California bearing ratio or CBR (a penetration test for the evaluation of the mechanical strength of road subgrades and base courses, with higher values representing harder surfaces) and is calculated using the following equation:

$$M_R = 2.555\,(\text{CBR}^{0.64}).$$

(4)

- The assumed specific gravities and costs of materials are listed in Table 7. Note that these costs are highly recommended to be updated regularly to reflect actual costs.

Table 9 provides a summary of all input parameters, default values, and assumptions.

Table 9: Ethanol and biomass worksheet input parameters

Parameter	Default value	Range	Comments
Plant capacity	User input	N/A	Input in MLY or MWe
Design period (years)	20	1 to 40	A service life of 20 years should be adequate for low-volume pavements.
Yearly growth factor (%)	0	0 to 100	This is capacity growth associated with the plant alone. It is not expected that a plant at full capacity will increase its capacity with time; thus, this value is set at 0%.
Ethanol plants only			
Percentage of corn used that is trucked to the plant (%)	100	0 to 100	The model assumes all corn is supplied to the plant by trucks.
Ethanol leaving the plant by truck (%)	20	0 to 100	The model assumes 20% ethanol is hauled from the plant in trucks.
Dried distillers grain leaving the plant by truck (%)	20	0 to 100	The model assumes most of the dried distiller grains leave the plant by rail.
Biomass plants only			
Capacity factor (%)	66.6	19 to 100	The average capacity factor of biomass plants across the USA is 66.6%.
Fuel type for biomass			

Corn stover (%)	25	0 to 100	Each plant uses different raw materials with different densities. This, in turn, affects the number of trucks supplying the materials. Currently, there is no dominant material in Indiana, thus the equal division.
Wood chips (%)	25	0 to 100	
Sawdust (%)	25	0 to 100	
Baled straw (%)	25	0 to 100	
Structural parameters			
California bear ratio	User input	N/A	CBR is then converted to resilient modulus (M_R) using Equation 4; a value of 3 or less can be used for conservative results.
Standard normal deviate (based on percent reliability), Z_R	75	50 to 99	The designer should choose the level of reliability of the design. For local county roads, the reliability is typically lower than high-volume roads.
Terminal serviceability, P_t	2	3.0 to 1.5	This is the value that reflects the condition of the pavement at the end of its service life. The AASHTO design guide recommends a value of 2 for county roads.
Standard deviation, S_o	0.5	0.3 to 0.5	This number reflects the variability within the pavements' materials. A value of 0.5 is recommended by AASHTO design guide.

AbdelWarith et al.

AbdelWarith et al. Energy, Sustainability and Society 2014 4:14

For ethanol plants, the number of trucks carrying products and raw material (or empty) is determined by multiplying the amount of ethanol produced per year (in MLY) by the number of trucks needed to ship one million liters of ethanol in a day (i.e., 0.25 for ethanol, and 0.0926 for the added value product DDG) or dividing by the average liters of ethanol carried by each truck daily (30,000 liters/truck). The weight of each truck is calculated using the weight of the truck empty plus the weight of the cargo (raw materials or products carried). The steering axle has a fixed load. Driver and trailer axles share the weight of the load equally. The loads are converted to load equivalency factors (LEF)ÿÿ¶5 using Equation 1.

In biomass plants, raw materials are burned to produce electricity. Each kilogram of raw material can produce a certain amount of heat energy. Typically, 1 kg of biomass can produce 2,746 W of heat energy; this is labeled as the production rate. The heat is then used to convert liquid water into steam, which in turn, rotates a steam turbine to produce electricity. The amount of heat needed to produce 1 KWh of electricity is the heat rate of the process. The heat rate is typically 4,103 W/KWh. The amount of material needed per day can be obtained by multiplying the heat rate by the number of KWh produced in a day and then dividing by the production rate.

The next step is to calculate the number of trucks needed to carry the raw material. First, the weight of each raw material is calculated; next, it is divided by its density and converted into volume. The total number of trucks is then calculated by dividing the total volume of that material by the capacity of each truck, which is typically 90.6 m^3. The weight of each truck can be obtained by multiplying the total truck capacity (90.6 m^3) by the density of the raw material used. The raw material used by the worksheet can be corn stover, woodchips, sawdust, and baled straw, or any combination of these. The loads are converted to LEF using Equation 1.

In both ethanol and biomass worksheets, the total number of ESAL is calculated by multiplying the LEF by the number of trucks in the design period that have that axle, and then by summing up all the ESAL.

The structural number (SN) for all layers is calculated using Equation 3. The thickness of each layer is calculated using the attained structural number, and the layer and drainage coefficients. The AASHTO [56] structural number equation is

$$SN = a_1 D_1 + a_2 D_2 m_2 + a_3 D_3 m_3. \tag{5}$$

For full-depth asphalt, the depth of the layer is obtained by dividing SN by a_1. The worksheet calculates various layer thickness combinations. The first layer is initially set to the minimum recommended by AASHTO [56] for the number of ESALs attained. The base and sub-base are calculated by satisfying two simultaneous equations. The first is the structural number equation, and the second is the ratio of base to sub-base thickness set by the user. The tool computes various combinations of the three layer thicknesses: D_1, D_2, and D_3. The user can chose any combination or change the values to produce a unique design. For full-depth asphalt, the depth of the layer is obtained by dividing SN by a_1.

The worksheet automatically estimates the costs for the thickness combinations using the assumed cost values listed earlier. However, the users have the option of specifying their own costs.

Wind Farm Design Specifications

Wind turbines have very large and heavy components. These components, when transferred to the wind farm location, can accelerate the deterioration of the road assets and their components. Similar to the ethanol and biomass procedure, there are two levels of input, primary and secondary. These are presented in Table 10.

Table 10: Primary and secondary inputs needed for the wind farm procedure

Primary inputs	Description	Secondary inputs	Description
Number of wind turbines	To transfer each wind turbine, a certain number of trucks are required, which typically use the same transfer routes. In the case that several turbines are transferred by different routes, each route should be designed for the number of turbines that will be moved across.	Tire contact area	Different tires have different contact areas. This value could be obtained from the manufacturer. The load and internal pressure are factors that affect and are affected by these values. If this value is unknown, 1,935 cm^2 is recommended as default.

California bearing ratio (CBR)	This value reflects the strength of the underlying soil. To get the exact value, soil bores are needed to be drilled in the construction location. However, the value can be closely estimated by knowing the type of soil in the area. Highway supervisors can resort to previous experience or soil maps to determine the soil type in the proximity.		

AbdelWarith et al.

AbdelWarith et al. Energy, Sustainability and Society 2014 4:14

To simplify the design process, a number of assumptions are made. It is important to note that all these assumptions could be changed manually in the design guide spreadsheets. The following are the assumptions considered in the wind farm design procedure:

- One-time heavy wheel loads during construction: Use Asphalt Institute's manual 'Thickness Design: Asphalt Pavements for Heavy Wheel Loads'. The loads necessary for a single wind turbine to be built (components and construction) are shown in Tables 4 to 6.
- Turbine loads based on a GE 1.5 s (1.5-MW design): Loads are assumed to be similar for all turbines near this size.
- The resilient modulus introduced in the design's internal calculations is calculated using Equation 4.
- The specific gravities and cost of materials assumed are listed in Tables 6 and 7.

Table 11 lists all input parameters and their default values, where applicable.

Table 11: Wind farm worksheet input parameters

Parameter	Default value	Comments
Number of turbines	User input	The number of wind turbines to be installed.
Soil CBR	User input	CBR is converted to resilient modulus (M_R) using Equation 4.
Tire contact area (cm²/tire)	1,935	This value could be obtained from the manufacturer. One thousand nine hundred square centimeters was used for dual-tire configurations [72, 73].
Maximum load per tire (kg)	4,536	Only construction loads are considered.

AbdelWarith et al.

AbdelWarith et al. Energy, Sustainability and Society 2014 4:14

Traffic counts are calculated by multiplying the heaviest truck used for design, by the number of trucks per wind turbine (assumed 91), and by the number of wind turbines (inserted by the users) to estimate the layer structural numbers, two values are needed to be calculated: the tire coefficient, a, and the tire pressure. The tire contact area coefficient, a, is calculated as follows:

$$a = \sqrt{(\text{Tire Contact Area}/\pi)} \tag{6}$$

Whereas the tire pressure, p, is calculated as

$$p = \frac{\text{Maximum Load}}{\text{Tire Contact Area}}. \tag{7}$$

The structural number is calculated as

$$SN = 0.3a\left(0.773\ln(p) - 2.535 + \frac{15 - M_R}{10.5}(0.049\ln(p) + 0.116)\right). \tag{8}$$

To estimate costs of various alternatives, the cost of the calculated thicknesses is computed using the assumed cost values listed in Table 7 (the worksheet does this computation automatically); however, the users have the option of including their own cost values.

Validation

The ethanol and biomass worksheets are based on the AASHTO pavement design guide [56], inheriting its strong points and its limitations. For the purposes of developing a user friendly design procedure, the AASHTO design method is preferred over the more recent mechanistic-empirical design method, as the former is well known for its empirical approach. Moreover, it does not need calibration or validation, because it is linked to the validated AASHTO design guide. For more conservative results, the reliability factor (ranges from 50% to 99.9%) can be increased in the worksheet.

Finding structural data associated with local roads is a tedious task. However, Jay, Posey, and Wabash counties did collect the pavement layer thickness values after they were upgraded for the construction and operation of the biomass and ethanol plants. For the ethanol plant in Jay County, provisions were made to accommodate the new increased truck traffic, and the road was resurfaced; however, the assumptions made related to the preferred truck-driver routes fall short, causing excessive deterioration to the adjacent roads. For the ethanol plant in Posey County, to handle the excess traffic, the county resurfaced all access roads. Similarly, to accommodate the anticipated increase in traffic associated with the ethanol plant in Wabash County, a 1.2-km road section was reconstructed. With these exceptions, the interviews with various county officials showed that there were generally no provisions made for ethanol and biomass plants due to their close proximity to state roads.

The second step in the validation process is to produce designs for these three counties using the developed worksheets. Table 12 presents the recommended pavement sections for each plant based on the ethanol and biomass worksheet, along with an estimated cost for each alternative. A CBR value of 3, representing tilled farmland, was used for Jay and Posey counties, and a value of 2 (representing softer surfaces) was used for the Wabash County. The county highway engineers mentioned that the soil is weak in that particular area. A design period of 20 years was assumed in all three cases.

Table 12: Ethanol and biomass worksheet results

Design	Alternative 1	Alternative 2	Alternative 3	Alternative 4
Jay County, POET Plant, capacity 65 MLY				
Surface layer (cm)	7.62	8.89	10.16	10.16
Base layer (cm)	26.67	15.24	30.48	19.05
Sub-base layer (cm)	34.29	40.64	15.24	30.48
Pavement cost ($)	246,744	247,468	262,624	259,417
Posey County, Abengoa Bioenergy Plant, capacity 88 MLY				
Surface layer (cm)	7.62	8.89	11.43	10.16
Base layer (cm)	26.67	15.24	30.48	21.59
Sub-base layer (cm)	35.56	44.45	15.24	30.48
Pavement cost ($)	249,060	254,417	282,769	266,306
Wabash County, POET Plant, capacity 65 MLY				
Surface layer (cm)	7.62	8.89	12.7	10.16
Base layer (cm)	29.21	15.24	30.48	25.4
Sub-base layer (cm)	38.1	49.53	15.24	30.48
Pavement cost ($)	260,082	263,182	302,414	276,160

AbdelWarith et al.

AbdelWarith et al. Energy, Sustainability and Society 2014 4:14

The final step in the validation process is to compare the pavements designed by the county engineers and private contractors, with the pavement designs proposed by the developed tool. Table 13 lists the capacities of various plants, and compares as-built thickness to those proposed by the worksheets. The structural numbers obtained by the worksheet are higher than the actual numbers from both cases. Wabash County pavements were designed by an engineering consulting firm.

The other counties used developer and local suggestions. The results of the worksheet are closest to the engineering firm's recommendations. With all inputs (except the plant capacity and the design period) being set at default values, this suggests that a proper design may be consistent with the worksheet's default output. If more information was available, such as the soil CBR or the design period, the results would likely be more accurate.

Table 13: Structural numbers of designed and actual upgraded pavements

County	Capacity (MLY)	Actual as-built pavement layer thickness (cm)			As-built structural number	Proposed structural number		
		Surface	Binder	Base		5-year design period	10-year design period	20-year design period
Jay	246	5.08	5.08	15.24 min	1.8	2.8	3.1	3.4
Posey	333	7.62	5.08	15.24	2.2	2.9	3.2	3.6
Wabash	246	10.16	10.16	17.78 stabilized	3.8	3.1	3.4	3.8

AbdelWarith et al.

AbdelWarith et al. Energy, Sustainability and Society 2014 4:14

Turning to the wind farm spreadsheet validation, it is noteworthy that in 2008, several lease agreements were signed with local farmers and land owners in White County, with the intention of building electricity-generating wind towers. A private consultant was hired to insure that the adjacent roads would be capable of handling the transportation of the wind tower components. The engineering firms performed an extensive field evaluation which included soil borings from all adjacent pavement sections, lab testing and soil classification of the collected soil samples, and performing non-destructive pavement testing using a falling weight deflectometer (FWD). Note that pavement surface deflection is typically used to evaluate the flexible pavement structure and the rigid pavement load transfer and is measured as the pavement surface's vertical deflected distance as a result of an applied static or dynamic load [79-82]. The FWD is the most common type of equipment to measure the surface deflection in Indiana, and the units used are thousandths of centimeters from a FWD center-of-load deflection, corrected to a 11,340-kg load applied on a 30-cm-diameter plate, adjusted for temperature (18°C) [83,84]. The engineering firms along with the White County highway department concluded that pavement upgrades were required. The engineering firm assumed the construction of 127 wind towers (phase I of the project) which included 5,000 concrete trucks, 8,000 gravel haul trucks, 1,150 semi-trucks for turbine component delivery, and numerous passes by medium and heavy cranes. This resulted in a total of 14,150 vehicles (plus crane passes). This value is more conservative than the 11,557 vehicles assumed by the wind farm worksheet. Even though the worksheet underestimates the number of trucks, it is important to note that the design methodology is based on the maximum truck weight for any truck category that constitutes more than 10% of the truck traffic. The consulting firm and the worksheets both used 4,536 kg per tire as their maximum weight.

The engineering consulting firm developed a pavement design that included the pavement layer thicknesses to carry the wind turbine components and construction materials. The SN was attained by the consultant using the AASHTO design guide for low-volume aggregate surfaced roads [56] and considered allowable rutting. Table 14 lists the consultant and worksheet results.

Table 14: Wind farm worksheet and county consultant inputs and results

Input criteria	Consultant design	Worksheet design
Number of turbines	127	127
Truck traffic assumed	14,150	11,557
California bearing ratio	3	3
Soil resilient modulus (Mpa)	41.37	35.58
Allowable rutting (cm)	5 to 7.5	N/A
Allowable loss of service	3	N/A
Aggregate base modulus (Mpa)	206.4	N/A
Percent heavy trucks (%)	80	100
Maximum load per tire (kg)	4,536	4,536
Recommended structural number	1.3	1.9
Recommended pavement layer thickness	-	-
Hot mix asphalt surface (cm)	0	5.08
Stabilized aggregate base (cm)	30.5	30.5

AbdelWarith et al.

AbdelWarith et al. Energy, Sustainability and Society 2014 4:14

As expected, the worksheet provides a higher SN than the one proposed by the consultant. This could be due to the fact that the consultant has more accurate information on this specific project due to the tests that were performed, or it could be due to the selection of design procedure. It can be argued that a low-volume road design is also applicable because the number of ESAL is expected to be small. However, the developed worksheet uses the heavy wheel design, which more closely matches the given traffic scenario (the heavy construction loads).

CONCLUSIONS

Data associated with sustainable energy facility traffic (such as number, type, and weight of trucks with or without cargo) were collected, to develop Excel-based tools (worksheets) and assist local agencies in the design of pavements in the proximity of ethanol plants, biomass plants, and wind farms. The worksheets provide a user-friendly environment for engineers with any level of expertise to produce a pavement design for the aforementioned facilities in an easy and timely fashion. Experienced designers have the option to change the default values of the worksheets in order to produce more cost-effective designs. Otherwise, the worksheets' default values can be maintained and still provide a conservative design.

From the comparison of the worksheet-generated designs and those practically implemented, it was found that the worksheet-proposed pavements were slightly thicker than the actual implemented designs, and thus less likely for the pavement to fail. This could reflect a need for collection of additional data points, or for further calibration of the tool through additional validation tasks. To that end, the as-built pavement sections will be revisited after 1- to 5-year intervals to assess their condition and further validate the worksheet tools.

The developed worksheets can serve as a hands-on tool to assist local government engineers in evaluating and quantifying the probable effects of the construction and operation of a sustainable energy facility in their jurisdiction. Further recommendations to assist in achieving this goal involve inclusion of biodiesel plants, further validation of the worksheets using measures of pavement distress (rutting or cracking), and comparison of the design outputs with actual data from constructed roads.

AUTHORS' CONTRIBUTIONS

KAW and WR collected the data and conducted the analysis. PA and KAW drafted the manuscript. JH, JF, and PA led and coordinated the study and affected its design. All authors read and approved the final manuscript.

ACKNOWLEDGEMENTS

The authors would like to thank Neal Carboneau and John Habermann for their useful comments. The contents of this paper reflect the views of the authors who are responsible for the facts and the accuracy of the information presented herein and do not necessarily reflect the official views or policies of the FHWA and INDOT nor do they constitute a standard, specification, or regulation.

REFERENCES

1. Bischoff A (2012) Insights to the internal sphere of influence of peasant family farms in using biogas plants as part of sustainable development in rural areas of Germany. Energ Sustain Soc 2:9

2. Dampier JE, Shahi C, Lemelin R, Luckai N (2013) From coal to wood thermoelectric energy production: a review and discussion of potential socio-economic impacts with implications for Northwestern Ontario, Canada. Energ Sustain Soc 3:11

3. Galich A, Marz L (2012) Alternative energy technologies as a cultural endeavor: a case study of hydrogen and fuel cell development in Germany. Energ Sustain Soc 2:2

4. Green JS, Geisken M (2013) socioeconomic impacts of wind farm development: a case study of Weatherford, Oklahoma. Energ Sustain Soc 3:2

5. Grunwald A, Rösch C (2011) Sustainability assessment of energy technologies: towards an integrative framework. Energ Sustain

Soc 1:3

6. Hagen Z (2012) A basic design for a multicriteria approach to efficient bioenergy production at regional level. Energ Sustain Soc 2:16

7. Halder P, Weckroth T, Mei Q, Pelkonen P (2012) Nonindustrial private forest owners' opinions to and awareness of energy wood market and forest-based bioenergy certification - results of a case study from Finnish Karelia. Energ Sustain Soc 2:19

8. Hassan MK, Halder P, Pelkonen P, Pappinen A (2013) Rural households' preferences and attitudes towards biomass fuels - results from a comprehensive field survey in Bangladesh. Energ Sustain Soc 3:24

9. Klagge B, Brocke T (2012) Decentralized electricity generation from renewable sources as a chance for local economic development: a qualitative study of two pioneer regions in Germany. Energ Sustain Soc 2:5

10. Mohr A, Bausch L (2013) Social sustainability in certification schemes for biofuel production: an explorative analysis against the background of land use constraints in Brazil. Energ Sustain Soc 3:6

11. Niemetz N, Kettl K-H (2012) Ecological and economic evaluation of biogas from intercrops. Energ Sustain Soc 2:18

12. Nishimura K (2012) Grassroots action for renewable energy: how did Ontario succeed in the implementation of a feed-in tariff system? Energ Sustain Soc 2:6

13. Oyedepo SO (2012) Energy and sustainable development in Nigeria: the way forward. Energ Sustain Soc 2:15

14. Palmas C, Abis E, and von Haaren C, Lovett A (2012) Renewables in residential development: an integrated GIS-based multicriteria approach for decentralized micro-renewable energy production in new settlement development: a case study of the eastern metropolitan area of Cagliari, Sardinia, Italy. Energ Sustain Soc 2:10

15. Scheer D, Konrad W, Scheel O (2013) Public evaluation of electricity technologies and future low-carbon portfolios in Germany and the USA. Energ Sustain Soc 3:8

16. Stoeglehner G, Niemetz N, Kettl K-H (2011) Spatial dimensions of sustainable energy systems: new visions for integrated spatial

and energy planning. Energ Sustain Soc 1:2

17. Taheripour F, Hertel TW, Liu J (2013) the role of irrigation in determining the global land use impacts of biofuels. Energ Sustain Soc 3:4

18. Taheripour F, Zhuang Q, Tyner WE, Lu X (2012) Biofuels, cropland expansion, and the extensive margin. Energ Sustain Soc 2:25

19. Tunç M, Pak R (2012) Impact of the clean development mechanism on wind energy investments in Turkey. Energ Sustain Soc 2:20

20. Walter K, Bosch S (2013) Intercontinental cross-linking of power supply - calculating an optimal power line corridor from North Africa to Central Europe. Energ Sustain Soc 3:14

21. Dooley F, Tyner W, Sinha KC, Quear J, Cox L, Cox M (2009) The impacts of biofuels on transportation and logistics in Indiana. Technical Report SPR-3133. Joint Transportation Research Program (JTRP), Indiana

22. Ginder R (2006) Potential infrastructure constraints on ethanol production in Iowa. http://www.econ.iastate.edu/sites/default/files/publications/papers/p3873-2007-07-27.pdf*webcite*. Accessed 2 May 2010

23. NADO (2007) Ethanol production impacts transportation system. National Association of Development Organizations Research Foundation Transportation Special Report 2:1-6

24. AP (2007) Road shuts ethanol plant. Associated Press. http://www.rapidcityjournal.com/news/state-and-regional/article_4edf0925-4fc1-5106-8d7a-12cadda26bad.html Accessed 3 Sept 2009

25. Wakeley HL, Griffin WM, Hendrickson C, Matthews HS (2008) Alternative transportation fuels: distribution infrastructure for hydrogen and ethanol in Iowa. ASCE 14:262-271

26. Kissel C, Cassady J (2008) Wind industry promises rural jobs, transportation challenges. http://66.132.139.69/uploads/nadort020608b.pdf Accessed 2 May 2010

27. Tidemann M (2010) Turbines for ethanol plant OK'd. Estherville Daily News. http://www.esthervilledailynews.com/page/content.detail/id/501320.html. Accessed 10 April 2010

28. Tanaka AM, Anastasopoulos PC, Carboneau N, Fricker JD, Habermann JA, Haddock JE (2012) Policy considerations for

construction of wind farms and biofuel plant facilities: a guide for local agencies. State Local Gov Rev 44(2):140-149

29. Reynolds RE (2000) the current fuel ethanol industry transportation, marketing, distribution, and technical considerations. CD-ROM. Accessed 3 Sept 2009

30. African Development Bank Group (2011) Updated environmental and social impact assessment summary: Lake Turkana Wind Power Project. African Development Bank Group, Kenya.

31. González J, Rodríguez Á, Mora J, Burgos Payán MM, Santos J (2011) Overall design optimization of wind farms. Renewable Energy 36(7):1973-1982

32. Kinoshita T, Ohki T, Yamagata Y (2010) Woody biomass supply potential for thermal power plants in Japan. Appl Energy 87(9):2923-2927

33. Kumar A, Cameron JB, Flynn PC (2003) Biomass power cost and optimum plant size in western Canada. Biomass Bioenergy 24(6):445-464

34. Ozerdem B, Ozer S, Tosun M (2006) Feasibility study of wind farms: a case study for Izmir, Turkey. J Wind Eng Ind Aerodyn 94(10):725-743

35. Tensar (2013) Retrieved March 7, 2014, from Wind farm access roads: two decades of floating roads. http://www.tensar.co.uk/~/media/548441FC4F6F4B6D90EB410B2969585B.ashx Accessed 7 March 2014

36. The British Wind Energy Association (1994) Best practice guidelines for wind energy development. The British Wind Energy Association, London.

37. Van Haaren R, Fthenakis V (2011) GIS-based wind farm site selection using spatial multi-criteria analysis (SMCA): evaluating the case for New York State. Renew Sustain Energy Rev 15(7):3332-3340

38. McAloon A, Taylor F, Yee W, Ibsen K, Wooley R (2000) Determining the cost of producing ethanol from corn starch and lignocellulosic feedstocks. Technical Report NREL/TP-580-28893. US Department of Agriculture and National Renewable Energy Laboratory, Golden.

39. Shurson J (2010) Distillers grain by-products in livestock and poultry feeds. http://www.ddgs.umn.edu/GenInfo/Overview/

index.htm Accessed 10 Oct 2010

40. AFDC (2009) Data, analysis and trends. Alternative Fuels & Advanced Vehicles Data Center. http://www.afdc.energy.gov/data/categories/vehicles. Accessed 15 Oct 2009

41. RFA (2008) Industry Statistics: US fuel ethanol demand. Renewable Fuel Association. http://www.in.gov/isda/biofuels/. Accessed 11 July 2010

42. Bruglieri M, Liberti L (2008) optimal running and planning of a biomass-based energy production process. Energy Policy 36:2430-2438

43. EIA (2007) Energy and economic impacts of implementing both a 25 percent rps and a 25 percent rfs by 2025. Energy Information Administration, US Department of Energy, Washington, DC. http://www.eia.doe.gov/oiaf/servicerpt/eeim/issues.html Accessed 11 July 2010

44. EIA (2009) Alternative fueling station total counts by state and fuel type. Energy Information Administration, US Department of Energy, Washington, DC. http://www.afdc.energy.gov/fuels/stations_counts.html Accessed 11 July 2010

45. CBEA (2003) the biomass power industry in the United States. California Biomass Energy Aliance. http://www.calbiomass.org/ Accessed 11 July 2010

46. DNR (2010) Woody biomass feedstock for the bioenergy and bioproducts industries. Indiana Department of Natural Resources. http://www.extension.purdue.edu/renewable-energy/docs/IBEWG/fo-WoodyBiomass_final.pdf*webcite*. Accessed 1 Feb 2011

47. RED (2009) Biomass Milltown Power Plant. Renewable Energy Development. Accessed 1 Feb 2011

48. Bastos CP (2010) Contributions of solar and wind energy to the world electrical energy demand. http://www.sefidvash.net/fbnr/pdfs/Solar_and_Wind_Energy.pdf Accessed 1 Feb 2011

49. Brown LR (2006) Wind energy demand booming: cost dropping below conventional sources marks key milestone in US shift to renewable energy. http://www.earth-policy.org/index.php?/plan_b_updates/2006/update52 Accessed 1 Feb 2011

50. EERE (2010) 80-meter wind maps and wind resource potential. http://www.windpoweringamerica.gov/wind_maps.asp Accessed

1 Feb 2011

51. Tchou J (2008) Wind energy in the United States: a spatial-economic analysis of wind power. http://www.gsd.harvard.edu/academic/fellowships/prizes/gisprize/ay07-08/Jeremy_Tchou.pdf*webcite*. Accessed 1 Feb 2011

52. AWEA (2009) US wind energy projects - Indiana. American Wind Energy Association, Washington, DC.

53. Lautal P, Stewart R, Handler R, Pouryousef H (2012) Michigan Economic Development Corporation Forestry Biofuel Statewide Collaboration Center. Task B1 Evaluation of Michigan Biomass Transportation Systems. Final Report. Michigan Tech Transportation Institute, Rain Transportation Program. p 112.http://www.michiganforestbiofuels.org/sites/default/files/Evaluation%20of%20Michigan%20Biomass%20Transportation%20Systems%20-%20FBSCC%20Task%20B1.pdf. Accessed 2 March 2013

54. McGregor B (2010) A reliable waterway system is important to agriculture. Technical report. Agriculture Marketing Service - US Department of Agriculture,http://www.ams.usda.gov/amsv1.0/getfile?ddocname=stelprdc5083396&acct=atpub*webcite*. Accessed 1 Feb 2011

55. Casavant K (2010) Study of rural transportation issues. Technical report, USDA and USDOT.

56. Denicof MR (2007) Ethanol transportation backgrounder: expansion of US corn-based ethanol from the agricultural transportation perspective. Technical report, United States Department of Agriculture.

57. AASHTO (1993) Guide for design of pavement structures. American Association of State Highway and Transportation Officials. Washington DC.

58. Asphalt Institute (AI) (1981) Thickness design—asphalt pavements for highways and streets. MS-1, Manual Series No. 1, 9th edn

59. AASHTO (2008) Mechanistic-Empirical Pavement Design Guide. Interim edition: a manual of practice. AASHTO, Washington, DC. p 212

60. Portland Cement Association (PCA) (2001) Thickness designs for soil-cement pavements. Portland Cement Association 30:EB068

61. Asphalt Institute (AI) (2001) Thickness design: asphalt pavements for heavy wheel loads. Manual Series No. 23 (MS-23). Asphalt Institute, Lexington.

62. Anastasopoulos PC (2007) Performance-based contracting for roadway maintenance operations. Thesis, Purdue University, West Lafayette, Indiana, M.Sc.

63. Anastasopoulos PC, Labi S, McCullouch BG (2009) Analyzing duration and prolongation of performance-based contracts using hazard-based duration and zero-inflated random parameters Poisson models. Transp Res Rec 2136:11-19

64. Anastasopoulos PC, McCullouch BG, Gkritza K, Mannering FL, Sinha KC (2010) Cost savings analysis of performance-based contracts for highway maintenance operations. ASCE Journal of Infrastructure Systems 16(4):251-263

65. Anastasopoulos PC, Florax RJGM, Labi S, Karlaftis MG (2010) Contracting in highway maintenance and rehabilitation: are spatial effects important? Transp Res Part A: Policy Pract 44:136-146

66. Anastasopoulos PC, Labi S, McCullouch BG, Karlaftis MG, Moavenzadeh F (2010) Influence of highway project characteristics on contract type selection: empirical assessment. ASCE Journal of Infrastructure Systems 16(4):323-333

67. Anastasopoulos PC, Islam M, Volovski M, Powell J, Labi S (2011) Comparative evaluation of public-private partnerships in roadway preservation. Transp Res Rec 2235:9-19

68. Anastasopoulos PC, Volovski M, Labi S (2013) Preservation: are 'public private partnerships' cutting costs? Pavement Preservation J 6(3):33-35

69. Anastasopoulos PC, Haddock JE, Peeta S (2013) Improving system wide sustainability in pavement preservation programming. J Transp Eng-ASCE 140(3):04013012

70. Wiltsee G, Wiltsee G (2000) Lessons learned from existing biomass power plants. Technical Report NREL/SR-570-26946. National Renewable Energy Laboratory (NREL). US Department of Energy, Washington, DC.

71. EPA (2006) National Electric Energy Data System. NEEDS. US Environmental Protect Agency. http://epa.gov/airmarkets/

progsregs/epa-ipm/BaseCase2006.html Accessed 20 April 2010

72. GE (2004) 1.5sl/1.5s Wind turbine. http://www.ewashtenaw.org/government/departments/planning_environment/planning/wind_power/Monthly%20Data_Reports/Attachment_1.pdf Accessed 7 Aug 2014

73. McKendry P (2002) Energy production from biomass. Part 1: overview of biomass. Bioresour Technol 83:37-46 PubMed Abstract |

74. Zuo Y, Maness P, Logan BE (2006) Electricity production from steam exploded corn stover biomass. Energy Fuels 20:1716-1721

75. Mani S, Tabil LG, Sokhansanj S (2004) Grinding performance and physical properties of wheat and barley straws, corn stover and switchgrass. Biomass Bioenergy 27:339-352

76. Armstrong J (2009) Wind farms and county roads. Interview performed by Wayne Richardson

77. The Manitowoc Company, Inc (2009) Manitowoc 16000 product guide. Manitowoc, Wisconsin.

78. The Manitowoc Company, Inc (2009) Manitowoc 999 product guide. Manitowoc, Wisconsin.

79. Fricker JD, Whitford RK (2004) Fundamentals of transportation engineering: a multimodal approach. Prentice Hall, New Jersey.

80. Till RD (2009) Overload truck wheel load distribution on bridge decks. Technical Report R-1529. Structural Section, Construction and Technology Division, Michigan Department of Transportation (MDOT), Lansing.

81. Eamon CD, Nowak AS (2003) LRFD calibration for wood bridges. CD-ROM

82. Anastasopoulos PC (2009) Infrastructure asset management: a case study on pavement rehabilitation. Ph.D. Dissertation. Purdue University, West Lafayette, Indiana. Available electronically from http://search.proquest.com/docview/304991168 Accessed 15 Aug 2011

83. Anastasopoulos PC, Labi S, Karlaftis MG, Mannering FL (2011) exploratory state-level empirical assessment of pavement performance. ASCE Journal of Infrastructure Systems 17(4):200-215

84. Anastasopoulos PC, Mannering FL, Haddock JE (2012) Random

parameters seemingly unrelated equations approach to the post-rehabilitation performance of pavements. ASCE Journal of Infrastructure Systems 18(3):176-182

Chapter 5

Identifying and Measuring Land-use and Proximity Conflicts: Methods and Identification

André Torre[1], Romain Melot[1], Habibullah Magsi[2],
Luc Bossuet[1], Anne Cadoret[3], Armelle Caron[4],
Ségolène Darly[5], Philippe Jeanneaux[4], Thierry
Kirat[6], Haï Vu Pham[7], and Orestes Kolokouris[8]

[1]UMR SAD-APT, INRA AgroParisTech, Paris Saclay University, Paris, France

[2]Department of Economics, Sindh Agriculture University, Sindh, Pakistan

[3]UMR Telemme, Aix-Marseille University, Aix-Marseille, France

[4]UMR Metafort, AgroParisTech, Clermont Ferrand, France

[5]University Paris VIII Vincennes Saint-Denis, Paris, France

[6]CNRS, University Dauphine, IRISSO, Dauphine, France

[7]UMR CESAER, INRA AgroSup, Dijon, France

[8]Panteion University, Athens, Greece

ABSTRACT

This text aims to present the methodology of study of land-use conflicts performed in recent years by a multidisciplinary team, and to reveal the methods of survey and data collection, as well as the structure of the resulting database. We first define the scope of our study by providing a definition of these conflicts, of their characteristics and motives, of the ways they manifest themselves and of the actors involved (I). We then present the methodology we have used to identify conflicts; it is based on a spatial analysis and the combined use of different data collection methods including surveys conducted by experts, analyses of the regional daily press and of data from the administrative litigation courts (II). Finally we present the resulting *Conflicts* © data base, with its tables and nomenclatures, in which the data collected in different fields are reconciled and analyzed (III), before providing a few examples of how this method can be used to analyze case studies in developed and developing countries (IV).

INTRODUCTION

Though conflict analysis is inscribed in a long tradition of social sciences, at the first rank of which lies sociology (Lewin, 1948; Touraine, 1978; Stephenson, 1981; Simmel, 2008; Freund, 1983; Coser, 1982; Wieviorka, 2005), researchers and practioners have preferred to focus their attention on the questions of conflict resolution rather than on the analysis of conflicts and of their characteristics (Castro and Nielsen 2001; Jeong 1999; Fisher, 1997; Neslund, 1990; Owen et al.,2000), except in cases of armed conflict (Boulding, 1962; Diehl, 1991; Hensel, 2001; Starr, 2005). Yet, the growing concerns about the environment, the issues of sustainable development, urban sprawl processes and about questions about people's living environment have recently led to renewed interest for issues related to land use conflict, also called land use and neighbourhood conflicts or environmental conflicts (See, among many others: Humphreys, 2005; Deininger and Castagnini, 2006; Magsi and Torre 2014; Mann and Jeanneaux, 2009; Campbell, et al., 2000; Darly and Torre 2013a, b; Cadoret 2009; Melé et al. 2004; Dziedzicki 2001; Charlier 1999; Cadene,1990).

Interest in these issues has grown in the fields of economics, geography, land planning as well as in sociology and social-psychology and has pointed to the necessity of analyzing conflicts, their occurrences, their impacts and main characteristics, more thoroughly. Such an approach obviously necessitates access to enough reliable data on conflicts per se, so as to be able to evaluate their number and volume, their role and impact, how they manifest themselves, what their causes or origins are and how they are solved.

Data about conflictuality is scarce and often incomplete for two main reasons. The first lies in the fact that little interest was shown in the subject until the years 2000. The second reason is related to the complexity of conflicts, which rules out the possibility of using only one representative variable. Indeed, land use conflicts find expression in various forms (tribunals, media coverage, violence…) which prevent one from making a simple representation and which explain why various disciplines are necessary to define them. A conflict that gives rise to analysis is a construct founded on information collected from different sources.

In light of this problem, a researcher wishing to study conflictuality must collect his/her own data (either they are focusing on developed or developing countries), and then analyze them (for more on the subject read Rucht and Neidhardt (1999) which describes the different stages that are necessary to this type of work). This is what we have done by developing a program of study on conflicts involving several teams of French researchers from various research institutes and universities. This program has successively focused on various issues pertaining of conflicts in natural, rural and peri-urban areas in developed and developing countries (France, Greece, Canada and Pakistan). Their contributions are based on analytical method adopted according to the nature and availability of data from different economies, and identification of land use conflicts, involvement of actors with different relations and network links.

This text aims to present the methodology of study of land-use conflicts performed in recent years by a multidisciplinary team, and to reveal the methods of survey and data collection, as well as the structure of the resulting database. The originality of this approach lies in its refusing to only use one particular source of information or one single innovative formula. The method we use to identify the conflicts, and which we present here, is complex and multi-dimensional. It involves a specific methodology that rests on the combination and triangulation of different sources and modes of data collection, and on precise protocols of data processing and of identification of conflict patterns, applied in each stage of the program. Following these protocols ensures that we achieve the most realistic representation possible of conflict within a given space or area. Our method borrows from other investigation procedures previously developed (See for example, Charlier (1999) based on press articles) and which we have transposed, improved and specified, but it also rests on innovative procedures (the use of litigation data and their treatment by means of software). It is based on social science analysis techniques (statistical inquiries, interviews, surveys, accounts, group follow ups…). It also consists in exploiting databases (such as the*Lamyline* database) or data provided by administrations (such as tribunals' judgments).

In some ways, our approach is not unlike that of other current projects of research on land use conflicts, in particular urban ones (see the research conducted in Canada (Trudelle, 2003; Joerin et al., 2005), in Netherlands (Leeuwen, 2010), or in Brazil (Observatorio Permanente dos Conflitos Urbanos na Cidade de Rio de Janeiro 2010). It is situated within a tradition that includes remarkable works conducted by teams of researchers located in various countries. Among these studies, three are, in our view, particularly significant, and have paved the way for our work by indicating the main difficulties involved in this type of research and by suggesting many solutions and possible paths for further research (Janelle, 1977; Ley and Mercer, 1980; Rucht and Neidhardt, 1999). Although we have not always drawn the same conclusions or adopted all the suggestions made by those authors, their works have been essential sources of inspiration. We owe a great deal to these studies in that they have helped us reflect on our multidimensional method of conflict analysis and avoid many pitfalls.

Two reasons explain why we have wished to present our method of analysis – which we have tested and improved over the last 7 years – to a wide audience of researchers and practitioners:

- We have wanted to show that it is possible to identify conflictual events and to deduce from them a general picture of conflictuality as well as a description of the characteristics of the conflicts that occur in a given area;

- We have also wanted to share our experience and encourage researchers to use our method, by responding to its associated criteria of intellectual property.

The structure of this article is as follows: we shall first present the scope of our investigations, placing particular emphasis on the questions of definition, origins and expression of conflicts. The second section of the paper is devoted to the presentation of our data collection method. We begin by describing the procedure of identification and diagnosis of the conflicts that occur in the area we have selected, and then discuss the identification method per se and more specifically its three foundation stones: the analysis of the daily regional press, of administrative litigations and the exploitation of surveys conducted by experts. In the third section we first present the *Conflict* ©database compiled from the information gathered. It addresses the questions of the construction and classification of the objects of conflicts, of the actors› profiles, of their utilization of land and of their arguments, successively. We end the article with a brief presentation of the results obtained from the exploitation of the database on conflictuality on selected case studies from developed to developing countries.

Definition of the Scope of Investigations

We are interested in land use conflicts and conflicts over resources. In order to identify them, it is, first, necessary to propose an operational definition that will enable us to both recognize and capture conflictual elements and situations, and to classify them in such a way as to be able to trace the conflictual profiles of a given area.

Our research studies are conducted in rural and peri-urban territories. They pertain to conflicts and tensions related to public consumption commodities (air, landscape amenities and nature's functions), resources (water or energy), waste and pollution, as well as to the areas of location of individuals or activities and their neighboring areas.

We call conflict an opposition marked by an engagement or a commitment between two or several parties (the actors of the conflict), in relation to local material objects. These oppositions reveal local characteristics related to spatial dimensions (e.g. topologic dimensions, neighborhood and transport infrastructures) as well as to social and economic characteristics linked to the areas in which they arise. Land use conflicts are the results of the dissatisfaction of one part of the population with actions undertaken or planned by their neighbors, by private institutions or by the public authorities. They are the pointers of the innovations taking place in the different territories and of the resistance they generate, and also the ferment of new innovation phases. We do not consider it necessary to eliminate conflicts nor even to try and solve them at all costs, for they are the expression of the voiced opposition of parties that consider themselves injured. Conflictual events are phases of coordination between actors and a way of reintroducing new actors in the mechanisms of decision making and of creation of territorial development projects. Conflicts can often be considered as part of a trial and error method regarding public decision: if one decision is considered opposite to the needs and wills of local populations, it could lead to tension, and afterwards to conflicts.

A Local Materiality

The conflicts we are interested in are distinguishable by their localized nature (i.e. territorial superposition of contradictory interests, rivalries between contiguous or neighboring areas), by the materiality of the objects that cause them or are concerned by them, as well as by the fact that they emerge in relation to differing land uses. Oppositions between individuals or groups pertain to concrete objects, to technical acts that are taking or will take place and imply concrete actions. These conflicts can have a strictly local component, or be related to questions that are more universal in scope. Whatever the initial situation, they can expand socially and spatially if they crystallize issues of a societal nature.

Land use conflicts have a territorial dimension. They rest on a physical basis: they take place between actors (sometimes, but not always neighbors) affected by a problem that has emerged and they develop around the use of localized support material or immaterial goods. They are inscribed in a geographic institutional framework, determined by both the actions of local and supra local authorities and by the rules they introduce. Indeed, *territoriality* and *unequal exposure* are central characteristics of land use conflicts. Environmental disputes are, first of all, *territorial* in nature to the extent that they involve the very unequal exposure of different territorial regions to environmental pollution, land speculation and development projects. Regions where the highest conflicts concentration is observed are also the regions where the densest concentration of risk-bearing facilities or building permits can be found.

Conflictual events are identifiable in relation to specific goods or pieces of land, i.e. the space within which the uses are in opposition. The cases studied in our research pertain to questions related to land, to territorial development as well as to water and its management, to the superposition of uses (production, tourism, leisure), to the development of economic industrial and port operation activities, to landscape and their evolution through urbanization and the creation of new equipment such as wind turbines, waste water treatment plants, waste management facilities, etc. More specifically, researches implemented in a country like France have been focused on conflicts dealing with the issue of urban sprawl and its consequences on speculation in farm land and forests (in particular risk of arson in peri-urban forests). In a Mediterranean country like Greece, forest fires can be mobilized as a strategy of land appropriation. Arsons can be used in speculative strategies to change land use destination, but can also indirectly provoke a local collective actions in favor of a stricter policy for the preservation of woodlands. Local inquiries paid also attention to the topic of farmland consumption and preservation of natural spaces.

The conflicts we have examined and their evolution pertain to the manifestations of oppositions to land modifying projects as well as to the effective emergence of constraints, pollutions and nuisances related to the changes that have occurred in the original space. Thus, the emergence of conflict is not necessarily related to the occurrence of a material event, but can also correspond to an expectation by certain categories of actors opposed to the project that that event will occur.

The Participants to Conflicts: Different Actors and Combination of Actors

The individuals or organizations involved in land use conflicts can be divided into two large categories:

Users of land and resources for productive purposes (whether or not they are the owners of the land in question and of their work tool): artisans and industrial entrepreneurs, farmers, herders and forest entrepreneurs, providers of recreational services involving the use of land;

- Users of land and resources for non-productive purposes (present on the land permanently - residents, hunters, sportspeople, hikers – or only present occasionally – tourists, secondary residents).

These categories of users find themselves involved in the tensions and conflicts, either individually or as parts of networks or groups of actors. Conflicts can involve users opposing one another (whether or not they pursue identical production goals), reveal oppositions between different categories of users. However, we consider that many actors can combine productive and non-productive functions, thus going beyond the simplistic dichotomy between the ones and the others. Indeed, they reveal contemporary social complexity and the various roles one person can play.

It is for this reason that we have chosen to base our work methodology on actors rather than on the uses they make of the areas considered. We also use the term "actor" so as to avoid referring to large categories of land users (residents, farmers, environmentalists, industries....) which are abstract and often account for only part of the reality and complexity of the actors themselves and of their relationships with others. Following the example of Janelle (1977) and of Ley and Mercer (1980) we also use the term "conflict participant" - which is the base economic and social unit of conflictuality - or conflict stakeholders.

The Motives of Conflict

Conflicts arise from changes or projects of change, perceived by some actors as contrary to their interests or their wishes. The material expression of changes at the origin of conflicts pertains to several categories:

Construction, deterioration or destruction of property, a landscape or infrastructure;

- Creation of a new production facility or expansion of an activity;
- Emission of external negative effects (diffusion pollution, odors, water drainage);
- Development of a property or piece of land;

- Access issues (restriction/exclusion, or opening/easement).

Nevertheless, the properties or facilities do not necessarily have to actually exist for a conflict to emerge. Conflicts can result from projects of construction, of implementation or extension of an activity, of development or of access modification, or from the emission of negative effects. In this case, conflicts are considered "anticipative" or "preventive". Indeed, disputes may arise between local residents or associations (acting individually or together), on the one hand, and the administration and applicants for permits to develop building projects or to operate activities, on the other. These disputes concern application for permits, above all, and come into play *a prior i*(preventive conflicts), i.e., before the environmental risk-bearing activity or development project is actually launched. In this case, the dispute arises at the outset of the process. However, disputes may also arise when industrial operators or developers challenge administrative decisions or when local residents protest against the consequences of pollutions, nuisances and environmental impacts and the dispute occurs then *a posteriori* (curative conflicts).

From Tensions to Conflicts

The distinction between tensions and conflicts is tricky to analyze. Indeed the emergence of a conflict follows an explicit engagement of the actors; an engagement which takes the form of acting out episode: threats, assault, actions at law, technical acts, and signs (forbidding access....). We shall call conflict any tension that turns into a declared confrontation, via the engagement of one or several parties.

We then consider that tension between various parties designates an opposition without the engagement of the protagonists, whereas a conflict emerges with the engagement of one of the parties. This engagement is defined by the implementation of a credible threat, which may take different forms:

- Legal actions
- Bringing the matter to the attention of the public authorities or of the civil service representatives
- Mediatisation (bringing the matter to the attention of the media, press, radio, television…)

- Assault or verbal confrontation
- Putting up signs (signs forbidding access, fences and gates…).

More or Less Apparent, Individual or Collective Manifestations of Conflictuality

Sporadic or recurrent, the tensions and conflicts related to the different land uses can manifest themselves in various ways:

- At inter-individual level: bad relations between neighbors, assault, recourse to third parties, retortion, reprisal;
- At a more general level: carried or handled by individuals (elected officials for example);
- Finally, at collective level, carried or handled by groups, particularly associations representing actors using land for non-productive purposes (these groups distinguish themselves from enterprises or large scale farmers or forest entrepreneurs in that they are characterized by a non-hierarchical internal organization and non-productive purposes), administrations, local or territorial authorities.

The strategies of the groups and individuals are closely related to the conflictual events. Highlighting them helps to explain the objectives and positioning of the actors at the onset of the conflict and in the ways to manage it. The way tension or a conflict is managed often depends on the goal pursued.

Tensions and conflicts have histories. Some conflicts pass away quickly, whereas others can last for long periods, with phases of more or less intense confrontation and stages of more or less latent antagonism. A phase of tension can last for a long time without turning into conflict, if the actors do not engage. During the period of tension preventive actions can be undertaken in order to prevent a conflict from happening. Nevertheless, it is important to note that it is perfectly possible for a conflict to arise without their having been prior tensions.

Modes of Prevention and Management of Conflicts

Although we make no hypothesis about the necessity to solve conflicts we regularly meet actors who seek to promote or implement modes of conflict resolution. The modes of conflict prevention and management can:

- consist of preventive actions aimed at promoting appeasement and at avoiding the occurrence of a conflict. These actions can be taken so as to promote inter-individual negotiation; they can consist in involving a third party representing land users, or in encouraging the actors to adopt a non-judicial route such as institutionalized mediation for example;
- merely consist of an arrangement between the actors involved;
- rest on a regulatory or legal techniques.

The Solutions Considered

Without going into an in depth description, we identify below the main types of solutions implemented:

- Technical acts;
- Compensations (financial, natural or technical);
- Land use planning;
- Eliminating the activity from the area in question or moving it somewhere else;
- Adjudication;
- Mediation by an insurance company.

Methods of Conflict Identification

The identification and analysis of conflicts rest essentially, in our method, on following sources of data collection:

- The daily regional press (DRP);
- Surveys conducted by experts;
- Data from administrative litigation courts;

- Data through other sources are often added in studies on developing countries, where some of the previous relevant sources are missing;
- Internet (personal blogs and web pages)
- Geographic Information System (GIS).

Every source of data collection has its own limits. Only a small part of the conflicts (e.g. regarding the engagement paradigm previously identified) is treated by tribunals and a lot of opposition does not go on Courts. As far as the political traditions and the laws are different in each country, tribunals can play different and more or less important roles. The judgment can also be subjected to various interpretation or contestation, even regarding the administrative component. In the same spirit, there are severe limitations to the use of DRP. First, it is obvious that media behave differently with regards to the countries but also the regions or the areas within a given country. For example, they do not behave the same way in the north or the south part of France because of local traditions of public expression. It is also true that some media lie or forbid crucial events, or even at least euphemize about certain events, in order to hide a part of them or to reduce their importance. The same applies to experts opinions: they can lie, cheat, euphemize, or sincerely forget some crucial events. For all these reasons, we have chosen to base our identification and our analysis of conflicts on the triangulation of three main sources of data. Each one has its own limits, but altogether they provide a credible image of a conflict, at least regarding to the above definition.

A conflict is a social construction, which is identified by the observer and depends on the definition of conflict events and the types of observations. In our method, the identification of the conflict is based on a triangulation of the main sources of information (DRP, Surveys, litigation data or other data). It is by cross checking and comparing the various sources of information that an evaluation of the state of conflictuality in a given area is made. This job is performed by the experts of the team, on the basis of the main data. The status of "conflict" bears on at least two identifications in the main corpuses of data, for example in the daily press and regarding verbatim by local experts. The types of conflicts identified and the comparisons between different data bases aim at identifying several discrepancies between local situations; for example, DRP helps in revealing early oppositions to local projects whereas administrative litigations are related to well informed situations and revelation of infrastructure projects by mainly public authorities.

The identification of the conflicts, is however, preceded by the identification and diagnosis of the study area. Indeed, we focus on precisely delimited geographical areas and we collect and analyze the data pertaining to conflicts occurring within those zones. This enables us to describe with precision the conflicts and their evolution, and limits the number of possible occurrences.

Identification and Diagnosis of the Study Area

The area examined is always situated within an institutionally determined zone. Its geographical delimitation rests on that of the local public institutions, such as: communities of municipalities, conurbation comities, counties, regional nature parks, sub-regional constituencies.... The study area can comprise one or several of these administrative zones – several constituencies for instance. We shall only examine the events that occur within the chosen area. An exception is made for litigation studies. Because the amount of data on the rulings of secondary tribunals is usually too small and therefore not representative, we examine the district or region in which the study area is situated.

Once the study area is defined we perform an area diagnosis which must enable us to identify its main social and economic characteristics and its salient features and actors involved within it.

The basic diagnosis report, of approximately 10 pages in order to make comparisons, must comprise:

- A general presentation (location, the geographic morphology, history, social and economic dimension…);
- A presentation of the activities that depend on the resources available on the territory;
- The main territorial governance structures, particularly the institutions.

The Daily Regional Press (DRP)

DRP reaches in the hands of about 400 million readers around the world every day. It is the second most popular medium after electronic media, and therefore an interesting tool of observing voices of regional population. Furthermore, it has the twofold specificity of being the main medium of local news delivery and, for each daily regional paper, to often enjoy a quasi-monopoly in their circulation area. The DRP articles are an easily accessible source of information on pre, during, and post-conflicts, and complements efficiently the data found in other sources, through surveys in particular. The DRP provides relatively detailed information about local affairs, which is not the case of national newspapers (Rucht and Neidhardt, 1999; Mc-Carthy et al. 1996).

The work consists in gathering the information provided by a given daily newspaper, by examining all the issues published over a given period, of at least one year. The newspapers can be consulted online or in paper from, depending on availability. In the case of online consultation, the inventory can be conducted via the dissemination server providing access to the digitized editorial articles published by the main daily national and regional newspapers. The automatic keyword based search method is only used when information is sought about a particular topic, and not when a general inventory of all land use conflicts is needed. All articles are then displayed one by one, before being selected or not as part of our corpus. The articles quoted in the following sections were selected using criteria that have enabled us to differentiate the situations of tension from situations of conflicts (See Section Definition of the scope of investigations).

When the information contained in one article enables the researcher to identify the action, engagement and connection of an actor, or when the article provides complementary information about a previously described conflictual situation, it is indexed in the corpus using different variables:

- Its title
- Its date of publication
- The issue of the newspaper
- The section and page in which it appeared

- A very short summary of the facts described

Whenever possible a copy of the article is made (otherwise the summary replaces the copy). The list and the copies of the articles enable us, in laboratory, to group together the articles relating to the same conflict.

This process is not aimed at describing exhaustively all the conflictual situations. Rather, we deal here with a specific type of event, which is the one written about and reported to the public through the press. This source has important biases which rules out using it by itself. The press omits some events; it can have a tendency to euphemize, dissimulate, it can be partisan or controlled by certain interests. However, using the press has for several years been recognized as a means – in the perspective of an exhaustive quantitative analysis of conflict – of obtaining the "most complete account of events for the widest sample of geographical or temporal units", as Olzak (1992).

Surveys Conducted by Experts

Interviews with experts provide information about the level of conflictuality. We aim to cross check this information with other sources, or to highlight conflictual events that might not have been brought to the attention of the courts or of the DRP. The interviews are aimed at identifying, in each zone, the dynamics of evolution of the rural and peri-urban areas concerned, at determining the types of conflicts and tensions relative to competing uses of rural spaces and at finding out the solutions implemented in terms of territorial governance.

We take great care of the fact that the experts interviewed have no private interest on land or other local resources involved in the conflicts, so that there are not personally affected by changes in land use for example. For the case studies quoted in next section the interviews were conducted with experts contacted beforehand by telephone, using a list of 40 to 50 people per study area. We choose to interview experts from various professional and associative fields so that a wide variety of opinions were represented. Each interview session (face to face) lasts between two and three hours. Each expert can be interviewed once, or several times when necessary (depending on the sufficiently required information). The respondents were interviewed with an open questionnaire so as to obtain as much information as possible, concerning conflicts and their evolution.

The questions may not directly be related to the conflicts. Indeed it has been observed that asking direct questions about conflicts generally leads the interviewees to refuse to answer. The interviewers present themselves as conducting surveys on local situations of governance, of actions and interactions between actors, sometimes with specializations depending on the interviews and the institutions they represent. The questions are always indirect: the interviewer must be trained to identify the conflictual elements of a situation. They must ensure that the questions in the grid presented below are all answered. The analysis of the conflicts is conducted later, in committee meetings.

This work enables us to gain a more thorough understanding of conflictual processes, to describe them and to identify their components:

- The materiality of the conflicts;
- The actors of the conflicts;
- The motives of the conflicts and how they emerge, which contribute to generating them;
- The manifestations of the conflicts, which imply various levels of symbolic or effective violence, the engagement of the actors ranging from petitioning to engaging in legal proceedings, or assault.

The 10-page report of the conflicts analyzed on the basis of the interviews with the experts lists the main conflicts that have occurred in the area considered, according to the interviews conducted. Quantification is impossible but the report provides insight into the perceptions of some of the key local actors on the conflictual process. A short one-page synthesis is provided with the report summarizing the main information obtained.

The interviews with experts cannot be our only source of information about conflicts for they present strong biases due to the method of analysis: the actors may have forgotten some elements; they may amplify or minimize, omit or lie about certain points. It is therefore absolutely necessary to complement them with other sources of data. Nevertheless, the interviews provide information that other sources cannot provide, and help us to gain insight into the dynamics of local alliances and oppositions, by enabling us to interact with the actors of the conflicts or with the observers of conflictual situations and of their development in the long term. We agree, here, with the conclusions of other studies about conflicts drawn on the basis of interviews with local actors, and more particularly with the conclusions of a study conducted for the World Bank by Deininger and Castagnini (2006) about the conflicts related to land property in Uganda.

Thus, we conduct semi-directive interviews.

Interview and Interview Analysis Guide

The interview guide indicates the different items of information that must be collected in view of the analysis and exploitation. It does not provide with a clear description of the questions that must be asked during the interview. The interviewer must ensure that all items are discussed before ending the interview and must write a report. The questions to be asked and answered are the following; the order of the items is only indicative.

- Location (precise or extended)
- Spatial support of the conflict (punctiform or linear)
- Activities and their restrictions of use (productive, residential, recreational, "nature", exclusion, network infrastructures, public facilities)
- Number of actors (or groups) involved
- Actors (or groups) (by analyzing their degree of organization); distinguishing the actors who have an institutionally recognized mediation functional
- Origin/trigger of the conflict
- The causes mentioned
- Relation to time (duration, frequency)
- Relation to space (evolution of the space of concern during the conflict, by distinguishing if necessary, the places of litigation, the places where the conflicts take place and the places that are evoked by the actors)
- Forms of expression: (a) Verbal expression, individual expression (threatening mail, altercation, intentional damage of property,

confiscation) collection expression (petitions, pamphlets, collective occupancy, marches, etc.). b) Departure (eviction, non-participation, passive refusal to obey a legal order or a summons, lasting refusal to participate. c) legal action (civil or administrative jurisdiction)

- Demands/claims declared or not by the protagonists (elimination or minimization of the nuisance – with or without a proposal of technical solution), demand for material or symbolic compensation, interruption of construction, etc.)
- Development (possible solution, the conflict continues, agreement, tribunals....)
- Public facilities possibly cited during the conflict (as a cause, as an element of context, as a solution)
- Types of arguments put forward:
- effects on the personal living environment, versus activities, versus natural environment, collective, on the health of the individual or of the group, on the personal versus collective costs
- on the principle: transgression of a locally accepted rule of a group, of an official law
- Evaluation of the behavior of the actors who are statutorily assigned to intervene as mediator, as guarantors of the rules

We refer here to the list of local experts. They are obviously not all present on all the territories, but the goal is to provide information about each main category of the preceding Box and to achieve a balanced representation

The Experts to be Contacted on Each Site

Local Public Institutions

- Local elected officials: mayors of municipalities and other elected representatives (General councilors…)
- Directors/managers of inter-municipal services/facilities
- Economic director of inter-municipal planning
- Chairman of an inter-municipal body or of specialized committees (environment, tourism, agriculture, commerce…)

Institutions for the Protection of the Environment or Organizations of Nature Users

- Regional Directorate for the Environment (environmental protection agency)
- Local associations for the protection of nature or for the environment (water, hikers or other outdoor sportspeople)
- Federations of hunters and fishermen and their local associations
- Environment and Energy Management Agencies

Forestry and Agriculture Related Organizations (Specific Attention is Paid to this Area of Activities, and Particularly on Agriculture)

- National/regional forests offices
- Federation of Forestry Owners
- Local Land development and management bodies
- Chambers of agriculture
- Planning departments and agricultural economics services

Federation of Farmers Unions

- Regional directorates (planning, economic services, agriculture and forestry)
- Farmers (in person)…
- Socio-professional representation
- Chamber of trade
- Chamber of commerce and industry
- Entrepreneurs clubs
- Infrastructure planners
- Water agency
- Agency for electrification and water management
- Agency for highway/motorway construction
- Land Development and Rural Settlement Associations
- Land acquisition and relocation committees
- Department of railways
- Local construction agencies

Other State services

- Sub-prefectures (general secretaries or attached functionaries)
- Economic services

- District Court clerk
- Other sources of information
- Local press journalists
- Notaries
- Judges…
- Police department

Analyses of Litigation Rulings

Administrative litigations should be collected from respective courts for statistical analyses. Therefore, statistical analysis of judiciary sources aims at examining the way in which legal rules are mobilized in land use conflicts, using a study of litigations between the parties involved in conflicts. This analysis targets a specific category of conflicts: those that have taken a specific path and have led to legal proceedings.

The inquiry on conflicts, through the quantitative study of litigation, is an empirical methodology that has its own limits and opportunities. As for the limits, the research in legal sociology usually concludes, regarding empirical data, that the resort to courts is always intrinsically marginal, may the phenomenon be observed in developed or developing countries. The successive social acts that consist to get informed on its rights, to choose a legal advice to solve the dispute and to go to the courts constitute different stages that select the people involved in conflicts. But at the same time, even though the actors reluctant to settlement and going to courts are a small part of the whole set, this tiny fringe is likely to be a mass phenomenon concerning a huge quantity of cases that can be studied statistically. Furthermore, the study of this small fringe of litigants can inform us about non judicial settlements: for example when cases are filed but dropped before they are adjudicated, it may be a sign that the resort to courts was just a mean to open a negotiation.

In some developing countries, this selective dimension of litigation is surely more acute. Cultural and financial barriers for the access to courts, the likelihood of corruption among the judges, the weak probability of implementation of court decisions, given the failure of public administration, are disincentives for a judicial strategy. Nevertheless, if the social phenomenon of litigation is quantitatively less important in those countries, our researches (conflict over water dam in Pakistan) and other empirical inquiries (for example, the world survey of Antonio Azuela for the Lincoln Institute about the conflicts over land evictions) (Azuela and Herrera-Martin, 2010) show that the dynamic of the process can be favorable to a development of judicial strategies. In that case, the research will focus more on evolutions that indicate an increase of legal consciousness than on the intrinsically low level of judicial disputes on the short term.

In France, disputes concerning activities which bear environmental risk or are likely to impact farmland or preserved natural spaces are adjudicated by Administrative Tribunals. These frequently involve claims initiated by associations and local residents against projects which have been granted permits by administrative authorities but whose impact is alleged to be harmful to the health and well-being of the local environment and community.

The various forms of actions at law have also been the objects of fruitful investigations in matters of infrastructure development or urban planning. The study of the practices in penal law in this field reveals the maneuvers possibilities that exist at all stages of the treatment of the offence, and the possible solutions: by the administrative authorities that detects the offence, by the prosecutor who decide to not prosecute, or by the judge who decides to pass an alternative sentence. They also highlight the diversity of the origins of the offence reports submitted to the legal system, depending on the priorities determined locally by the administrations concerned – targeted prosecution campaigns over a given period so as to solve a local problem – in a field where offences are much more often reported by booking officers than by the victims of the offences (Struillou, 2004).

With the exception of rare studies targeting specific associative actors, (Leost, 1998), the statistical studies undertaken on the basis of data from administrative jurisdictions are almost non-existent, in the field of land-use litigation. These sources are only used in the framework of general statistical studies on the activities of administrative tribunals, which, however, provide a useful basis for conducting more targeted surveys (Barré et al. 2006).

Therefore, the analyses of conflicts based on the observation of legal and administrative litigations are conducted at small administrative unit (*department, district, division etc.*) level. This choice is justified by two arguments: on the one hand, that particular administrative unit is the territorial unit of reference for many actors, whether they be public actors in charge of land use regulation (prefect, non-central government services), or para-governmental and private actors: Associations for the protection of the environment generally act at these administrative levels; the same applies to chambers of agriculture or the associations of hunters and fishermen; on the other hand, court rulings generally state precisely where the problem is situated at municipal level (private litigations, applications for annulment of municipal decrees or deliberations) or at departmental level (applications for the annulment of prefectoral decrees).

Land use and environmental conflicts consequently reflect the way this administrative activity operates. Knowledge of the inner workings of this system on a normative and empirical level is imperative in order to understand the logic behind the litigation. In the case of administrative justice, this prerequisite seems easier to meet than in other fields, like criminal sociology, or the sociology of civil disputes, to the extent that the dispute can be traced to administrative decisions and activity for which there are more likely to be written entries. This may not be the case for private law documents or contracts, so it is easier to obtain data on the frequency of claims in the case of administrative justice. It is therefore with regard to the intensity of conflictual activity must be assessed.

In developed countries, especially in France that the corpus of court rulings is built using the *Lamyline* textual legal database which comprises the complete texts of the rulings made by the Court of appeals and Supreme Courts. More precisely, this database comprises all the rulings made by:

The State Council (highest administrative jurisdiction in France) since October 1st 1964,

- The Administrative Appeal courts since January 1st 1989,
- The Court of cassation (highest civil jurisdiction in France) since October 1st 1959 (excluding the rulings by the Criminal Chamber which are included from January 1st 1970 onwards),
- The Court of Appeal since January 1st 1982.

The software has a search engine that enables us to use the Boolean operators and those of several French case law data bases. The search for rulings by the four levels of jurisdiction defined above is conducted from the following case law data bases:

- State Council
- Administrative Appeal Courts
- the Court of cassation
- Appeal Court

Contrary, in the developing countries it is often difficult to collect the statistics from administrative courts, because of the absence of case law databases. Thus cases of the courts can be collected by searching manually from the yearbook records in their libraries, which is of course a time taking process. While in some developing countries uncooperative behavior of the administrative courts is discouraging the researchers[a]. Although the case law hierarchy is not the same in the entire developing world, it follows more or less the following structure:

- Supreme court
- High court
- Session court

Obviously, only part of the land use conflicts has been analyzed: those that have been dealt with by Courts of law, which limits the number of conflicts selected and examined. Indeed, the "passage through court" is the result of a selection using various filters: an individual's refusal to negotiate, an administrative authority's refuse to regularize... or on the contrary the intention to use the tribunal as a lever for starting stalled negotiations or for conducting them with a more favorable balance of power. These elements contribute to giving the conflicts that have been dealt with in court a profile whose specificity must be discussed in the interpretation of the surveys' results, without however considering that they represent all the conflicts that occur in the various territories.

[a]For example, in Pakistan tribunals refused to provide data on land use conflict related cases because of involvement of bureaucrats, feudal and politicians.

Data from Other Sources

The identification and listing of the conflicts are performed on the basis of the data from the three sources described above and on their comparison.

Nevertheless, other secondary methods of analysis can be added to this initial list, especially in developing countries or less developed countries, where some of the previous sources of data are missing. This can be particularly true for litigation data. These additional methods are aimed at conducting a more precise analysis or at specifying such or such dimensions of conflicts.

We shall, here, content ourselves with mentioning them. They are respectively:

- Sectoral analyses: for example, analyses targeting the farming sector or land property issues;
- In depth analyses carried out by researchers/experts in specific disciplines: for example, interviews of actors by sociologists, or monitoring of meetings or postal questionnaires administered in the context of research in social psychology;
- Study of particular situations: for example, study concerning a river, or farming facilities or land takes for road and water reservoir construction.
- In depth analyses of digital aerial photographs and remote sensing images through Geographic Informative System (GIS) of the territory; for example if the changes over a piece of land are being monitored in case of superposition of uses and its impacts;
- Analyzing personal blogs and web pages through internet, which is growing source of information on opinions and expertises in the developing countries.

Inquiries implemented in Greece combined different sources, from local newspapers to online websites supported by environmentalists associations ("observatory of natural spaces"). The latest source has the advantage to cover a large period (2001–2011), but is characterized by a selective approach, since data are furnished by local activists. Other sources complete the research methodology as follows: interviews with local experts (journalists, lawyers…) and key stakeholders like planning agencies, political leaders and elected officials, additional events collected through other web platforms (blog sites of environmental activists). An important source of documentations was also constituted by political materials like programs of local parties for municipal elections.

For analytical methods of the case study from Pakistan (presented in next section), data have been collected from various secondary sources to crosscheck the results. For example, the information has been collected from literature published by public and private organizations, on the case study perspective. The data have also been collected through aerial photographs and remote sending images, through satellites which were treated under GIS in order to examine land cover changes and impacts of the superposition of land uses due to decision for infrastructural projects by the public authorities over existed economic activities in the region. This technique of research became reliable after development in technical software. Moreover, data also collected after in-depth analyses from personal blogs and different web pages, these sources of information are very promising for future researchers of land use conflicts due to huge development in internet in the developing countries.

The Conflict © Database

The scope of our method is rather large, and we intend to provide it as an opportunity to studying land and resources conflicts in developed as well as developing countries. However, we have built a data base related to our own investigations, especially in France (13 case studies in different areas) and Pakistan. This data base can be expanded or reproduced for other types of investigations, in other countries or regional areas, given its wide scope and the general character of the contained items. We will present it in in broader terms in the following paragraphs.

In the context of the research conducted on conflict, using the relational database responds to a twofold analytical need. The aim, first of all, is to create data that can be exploited quantitatively, using the documents and survey results: the coding operations performed on the basis of the source documents are aimed at quantifying the phenomena of conflictuality, which will be analyzed and possibly examined from the perspective of the profiles of the areas concerned. But this database must also enable us to conduct a comparative evaluation: comparison between the sources used (court cases, newspaper articles, questionnaires), and comparison between the different areas surveyed.

The comparison between the different sources rests on the database' structure, which is designed in such a way as to be able to compare them. Besides the variables relative to specific observation contexts (type of rules used in the cases, number of press articles discussing a particular subject) some common variables have also been defined. The researchers conducting the coding operations must translate the specificity of their material in these cross sectional categories – described in detail below – relative to the types of conflicts observed, to the actors involved, or to the uses and arguments discussed.

Database Structure and Categories of Conflicts

The overall structure of the database can be schematized in the Figure 1 below. The data base incorporates three main data tables, that is, in order of inclusion:

- A table containing the variables relative to the geographic locations of the conflicts (in relation to a municipality, a community of municipalities, a *département* or a district);
- A table indicating the variables describing the conflicts per se, that is, the cross sectional categories – which are identical whatever the source of the data, and the categories relative to a context of observation (The legal categories defining, for example, the nature of a request made to a jurisdiction);
- And finally a table providing information about the profile of the actors involved.

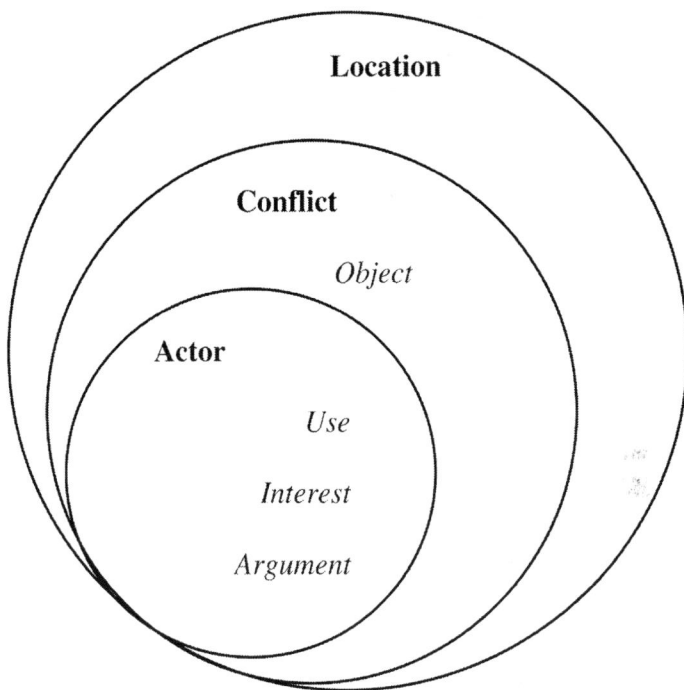

Figure 1: Simplified graph of the database' structure. The circles represent the relationships of inclusion between the tables. We have noted in italics the tables that are dependent on the "conflict" and "actor" tables.

Four other dependent tables must be mentioned: the description of the conflicts' "object", as well as information about the actors concerned: what the actor involved in the conflict uses the land for, the arguments the actor uses, and the interests that motivate the latter's engagement in the conflict.

The difficulty of studying conflict in one territory using different types of sources has to do with the identification of comparable analytical categories, whatever the observation point selected. This difficulty is crucial in the analysis of the requests sent to tribunals, in so far as those requests are expressed in a constrained language, that of the legal categories. Moreover, the enumeration of the criteria defining the contours of a conflict as a homogenous entity is a methodological demand which becomes particularly acute when exploiting press articles. Thus, while "formatting" the actions at law as clearly identifiable cases naturally leads us to choose "case" as the unit of measurement of the "conflictual" event, it is the "article" that is used as the measurement unit for the analysis of the press, when reconstituting the event after it has occurred.

Furthermore, a common definition of categories that cross the analysis of actions at law and that of the press, has to be found in order to enable the researchers to use a common language of description of conflicts. This does not exclude the necessity to define, for each type of observation material, specific variables relative to the source used. Thus, the nomenclature of requests, used by administrative court clerks for example, has been used for internally describing the different types of actions at law. Similarly, some variables are only meant to be used in the framework of an analysis of the press (number of articles devoted to a particular conflict). We present here the main components of the database, which is described comprehensively in several internal documents (for example, Galman and the participants of Conflict Program 2007).

The trickiest variables to define are, naturally, those that pertain to the different objects of conflict (see below). The detailed examination of the categories defined reveals the combination of several modalities of definition; complexity inherent to the expression of the forms of conflictuality. Indeed, the "objects of conflict" refer, depending to the cases, to:

- more or less clearly territorialized economic activities, whether they specifically involve using natural resources (agriculture, extraction of underground resources), location of production activities (industrial production, production of energy, waste treatment), or activities related to the presence of amenities (tourism);
- types of legal authorizations granted by administrative authorities, in as much as they correspond to land uses as defined by law;
- forms of social relationships marked by geographical specificities: Neighborhood relationships for instance.

The Objects of Conflicts: The Categories Used

- Accessibility and easement
- Right of access and passage

- Occupancy/parking
- Facilities classified for the protection of the environment
- Quarries, borrow pits
- Salvage/recycling, storing, waste treatment
- Salvage/recycling storage of materials (cars, tires…)
- Production, storage of chemical substances
- Other regulated industries
- Site rehabilitation
- Mining site
- Industrial production site
- Storage site
- Service activities
- Tourism, leisure
- Transport, Fuel distribution
- Trade, retail, advertising
- Road, rail, sea, air transport
- *Agriculture, forestry, fishing*
- Farming
- Forestry
- Fishing
- Public utility infrastructure
- Airport, railway, road, harbor infrastructure
  Construction of water reservoir
- Production – energy transport
- Telecommunication infrastructure
- Public facility
- *Management and Conservation of Natural Areas*
- Hunting/fishing
- Water quality management
- Soil quality management
- Air quality management
- Landscape, wetlands
- Conservation/ management of the Fauna/flora/biodiversity

- *Urban planning/development operation or document*
- Camping sites
- Urban development document
- Right of pre-emption
- Public domain occupancy
- Land restructuring
- Risk management areas
- Natural areas
- Management of natural areas
- Construction/extension of agricultural buildings
- Construction/extension of residential building
- Construction/extension of commercial building
- Construction/extension of thoroughfares
- *Neighborhood*
- Neighborhood nuisances
- Declaration of co-ownership
- Theft, damage, attack

Profiling the Actors Engaged in Conflicts: Categories of Actors, Land Uses, Arguments

The analysis of the actors engaged in conflicts is summarized with great care in the grid of the variables used. Once the credible engagement of an actor has been identified, one can, indeed, distinguish the actors that are at the origin of this action (the protesting actors) from those which this action targets (the disputed actors).

As stated in point 1.2., the typology of the groups of actors that we have used for the database is founded on a distinction between users of land and resources for production purposes, whether or not they are the owners of the land (farmers, forest entrepreneurs, artisans, industrial entrepreneurs, providers of recreational services…) and users of land and resources for non-productive purposes, whether they use the land in question occasionally or continuously (residents, hunters, fishermen, sportspeople, hikers, tourists, secondary residents…) (See below). If a local authority body, or a State department is, for example, user of resources (in the case of well boring for water for municipal consumption, or of public road maintenance by the public services), we shall consider the actor concerned as a "representative", considering that during a conflictual confrontation, s/he will refer to the authority which his role as representative confers upon him/her, and as defender of collective interests.

The Categories of Actors Involved in the Conflicts Studied

- Farming or equivalent occupation
- Farmers
- Irrigation operators
- Forest entrepreneurs
- Fish farmers
- Herders
- Industrial actors
- Artisans
- Industrial entrepreneur (mining activities)
- Construction and civil engineering firms
- Industrial entrepreneurs in the manufacturing industry
- Actors of the market service sector
- Transport enterprises
- Energy suppliers
- Providers of services related to waste management
- Providers of services related to water management
- Real estate developers, planners
- Actors of the hotel industry

Other market services

- *Associations*
- Ass. For the protection of the environment (national or worldwide)
- Ass for the protection of the local environment
- Hunters and Fishermen's Association
- Political party

- Other associations
- *Local public authorities*
- Region
- Department
- Municipality
- Public Intermunicipal Cooperation Institute
- Regional Natural Parks
- Local Administrative body (water agency, Committee for the protection of natural sites and monuments)
- *National Public authority*
- Minister
- Prefect
- De-concentrated State services
- Judicial authority
- *Elected representatives*
- Municipal, departmental, regional representatives
- *Professional organization*
- Trade unions
- Consular chambers
- *Individuals*

    Permanent residents

    Temporary residents

    Owners syndicates

    Tourists, non-sedentary population

Once the actors of the conflicts are identified, we describe the controversial or conflict generating land uses that is the land uses the spatial consequences of which the objecting parties consider as sources of constraints. The distinction between actors and uses requires the taking into account, through standard observation, of the fact that one user can use land for several purposes: A farmer can also be a herder, a hunter, or a defender of nature; an industrial entrepreneur may also be a hiker; a resident may in the framework of his/her work conduct polluting activities…

Different configurations of opposition emerge, including between actors making the same use of land. Particular attention is placed on the local public mechanisms that are liable to exacerbate or crystallize certain tensions into conflicts.

The Categories of Land Uses Identified for the Analysis of Conflicts

- *Creation of infrastructure*
- Production and transport of energy
- Road, railway, airport infrastructure
- Waterway transport infrastructure
- Waste management
- Water supply
- Public buildings
- Telecommunication
- Advertising infrastructures
- Recreational/tourism infrastructures
- *Production of services and exploitation of infrastructures*
- Tourism, hotel and catering industry
- Telecommunication
- Air, road, railway, river, sea transport
- Production and transport of energy
- Trade, advertising
- Sanitation
- Domestic waste transportation and management
- Water catchment, treatment, storage and supply
- Management of hazardous waste and materials
- Utilization of buildings housing public services (functioning)
- Exploitation of recreational or tourism infrastructure
- *Agricultural, fishing and forest production*
- Agriculture
- Aquaculture
- Breeding

- Professional fishing
- Forestry
- *Industrial production*
- Mining
- Manufacturing
- Site rehabilitation
- *Recreational and touristic use*
- Hunting, fishing
- Hiking, motor sports, non motor sports
- Tourism's events (cultural, musical, etc.)
- *Residential use*
- Construction/extension of social housing
- Development of poor housing
- Residential use
- *Conservation and management of resources*
- Fauna, flora, biodiversity
- Surface and underground water
- Site, landscape
- Ground
- Heritage
- Risks

Absence of a Characterized Land Use

Just like the distinction between actors and users, the identification of specific registers of argumentation is based on the hypothesis that some categories of actors can, depending on the conflictual situation, use specific arguments. The capacity for technical and legal argumentation can for example depend on the degree of collective mobilization which a given conflict gives rise to.

An important cornerstone for the analysis of conflicts consists, indeed, in highlighting the registers of argumentation used by the actors. Argumentative openness implies that the (individual or institutional) actors diversity their registers both from the point of view of values and that of rules, so as to lend as much efficiency as possible to their protest movement. It can perfectly well result in contradictions when the same actor puts forward both the individual's interest and the general interest. But beyond this classic opposition, these argumentative strategies appear to be more complex when using the "general interest" argument requires the mobilization of distinct norms, which are often revealing of contradictions in the law itself (Lascoumes, 1995): thus, in the context of a protest against an urbanization project, the principle of controlled development of land and the argument in favor of housing development are used in turns; both categories of general interest are recognized by legislation.

Registers of Argumentation: Categories of Analysis

- Scientific and technical argumentation
- Protection of ecosystems
- Infrastructures and equipment
- Risk evaluation and management
- Socio-economic argumentation
- Economic interest
- Sustainable development
- General local, regional or national interest
- *Reference to private land rights*
- Access
- Ownership
- *Responsibility*
- Private responsibility
- Public responsibility
- *Quality of life*
- Perception of risks
- Living environment
- Insecurity of individuals and property
- *Values*
- Modernity
- Tradition

- *Respect of the law and regulations*
- Risk prevention
- Other

Constitution and Exploitation of the Data Originating From the Daily Regional Press

After collecting the data from the DRP, the inscription is made in the data base using the following schemes. For each conflict identified, we include in the database a series of descriptive variables that groups together the following elements:

- A brief summary of how the conflict developed, as described in the relevant article or series of articles,
- The spatial environment of conflictual events, the facilities - being used or subject to usage restrictions – that are blamed as sources of constraints,
- The resources that have been modified, and the usage or activities affected by these modifications (reconstituted using the provided arguments),
- The opposing actors and their modes of engagement in the conflict,
- The geographic location of the goods/resources affected by the controversial facilities[b].

[b]One can associate to each conflict the list of municipalities in which are located the sites that are at the center of the conflict.

After analysis, we compile a 10-page long data sheet about the conflicts reported in the DRP. It lists the main conflicts that have occurred in the zone considered and classifies them according to the number of times they have been mentioned in the press. This helps us to obtain a picture of conflict, and of the media impact of the different causes of conflicts. This datasheet is accompanied by a one page synthesis summarizing the main information obtained.

Constitution and Exploitation of the Data Originating from Litigation Rulings

Once the database originating from litigation court ruling is compiled, it is analyzed statistically and lexically. The decisions are coded in such a way as to constitute a database integrated, first, into an Excel spreadsheet and then incorporated into a data processing software system (4D). The variables and modalities are defined using the above mentioned grid of analysis of conflicts also used for the interviews with experts and the DRP.

That corpus is exploited through the sequential and cross comparison of the significant variables, and through the analysis of the frequency of the legal references made. This analysis is completed by the textual statistical analyses performed with the software ALCESTE, so as to identify, through the language used in courts, the local specificities of the conflicts. These lexico-metric tools have only been used in the framework of the analysis of legal documents, which are well suited to this type of analysis due to the highly formalized structure of the argumentation. Without describing in detail the modalities implemented and the results obtained with this tool (Kirat and Torre 2004), it is useful to specify its scope as a component of a wider methodological approach to conflict analysis. The analysis of lexical statistics is justified by the fact that the language used in the court rulings is the product of the legislative or regulation texts used by the parties or the judge; this terminology is used to describe objects of protest or demand but also to express logics of action and goals (obtain the re-establishment of a right-of-way easement, the cancellation of a building permit in a zone of ecological interest, or the cancellation of a public inquiry conducted in the framework of a project of creation of a Classified facility for the protection of the environment). The lexical analysis is one of the elements of the study of the arguments mobilized by the actors. It enables us to show that one generic type of conflictual factors does not give rise to one unique and universal model of action in administrative courts. In other words, the actors seeking justice do not behave in a homogenous manner in all the *cases* studied.

The use of ALCESTE is not a technique for a-priori hypothesis testing, but for exploration and description. The program generates an empirically based classification of text units according to the pattern of co-occurrences of word tokens within these units. The lexical analysis software operates by dividing the corpus in elementary context, that is to say, a group of words identified by their length and punctuation. The similarity of these sentences is determined by reference to the similarity of the words used. A descending hierarchical classification distinguishes several classes of context units, and measures the closeness or distance between them. The classes identified constitute a basis for a principal component analysis (PCA), which is then used to realize graphical projection structured by the two most significant vectors. The researcher then begins to make interpretations based on the output from the program. It facilitates the discovering of the salient meaning structures and implements mechanisms for an independent analysis of the meaning of words, in order to get a statistical ranking of lexical statements in a given corpus.

The search for court rulings is performed by crossing the name of the departments selected with several keywords, defined in such a way as to cover as comprehensively as possible the scope of legal questions in which the land use conflicts can be formulated.

The Search Keywords Used for the Compilation of a Sample of Cases

- Easement
- Hunting or hunting right
- Building law
- Farm laws and hunting
- Farm laws and protection of nature
- Co-ownership or nuisance
- Directive 92/43/EEC
- Land application or breading
- Fauna and flora
- Classified facility
- Disturbance
- Nuisance and olfactory
- Nuisance and noise/sound
- National Natural Park or Regional Natural Park

- Passage and (hiking or Moto or quad)
- Rural land reparcelling
- Normal nuisance or neighbourhood nuisance
- Wetlands or swamp or bog

The judgments made in cases that occur locally at departmental level can be identified thanks to the effect of standardization of court decisions: the latter must include the address of the parties involved in the trial. Moreover, most decisions include the name of the study zone.

For example, the text of a judgment rendered by the French administrative litigation court is systematically made of four sections. The first states the identity of the petitioners, the nature and date of issue of the contested administrative act, and the administration that has issued the act, which is therefore sued in the litigation. The following section describes the judge's answer as to the means of form used by the different parties, that is to say the arguments relative to elements of the procedure (receivability of the request, the petitioners' right to act, etc.). The third section contains the judge's answer concerning the substance, that is to say the arguments used to oppose or to defend the administrative act itself. The fourth and final section evaluates the sanctions and compensations to impose depending on the ruling given.

At the end of this search, the corpus is reviewed so as to eliminate duplicates; indeed strong redundancy between the decisions found using different keywords can occur. The non-relevant decisions are subsequently eliminated, such as for example, the judgments concerning hunting accidents, or those relative to cases occurring in departments other than that where the parties reside...

Once the data is analyzed a 10-page datasheet on the conflicts identified through the exploitation of the litigation courts' data is compiled. It describes in detail, for each type of jurisdiction concerned (administrative and civil), the categories of requests that are the most frequently submitted to tribunals, and also indicates the main argumentation strategies observed according to the categories of actors (what corpus of rules is mobilized by what type of actor?). The question of the "outcome" of the conflicts, sometimes difficult to evaluate when analyzing the press, is here systematically interpreted, in so far as with the exception of abandonment or discontinuance of a procedure, a case judged on the basis of its substance gives rise to a decision that will oppose a "winning" party to a "losing" party. Thus analysis of the "rate of success" per category of conflict and per type of actors is therefore an important component of this synthesis. As a matter of fact, as in the case of the press and the interviews with experts, a one-page synthesis summarizes the main information obtained.

Added Local Socio-economic Data

In addition to these elements we include in the database information relative to the area and municipality concerned and useful for understanding the local context in which conflicts emerge. These data are of two types:

- First of all, they are socio-economic variables describing the profile of a territory from the point of view of social dimensions (tax related data, proportion of social housing...), of environmental issues (percentage of farm and natural lands, of areas protected for their heritage value), of the demographic dynamics (migration, population pyramid), etc.

- A second group of data provides information about the different administrative decisions that are liable to give rise to opposition: Building permits granted by mayors (data on the authorizations granted by the Regional Directorates of Infrastructure), or permits issued by the prefectures in accordance with regulations *concerning classified facilities.*

Both types of data refer to two levels of explanation of conflictuality: The first, immediate, level of explanation helps us to evaluate a rate of conflictuality in relation to an activity of reference that generates controversy. Thus, intense conflictuality concerning urban development generally reflects a highly dynamic construction market. The succession of protests related to pollution issues is often the consequence of an area being highly exposed to nuisances because of the important number of classified facilities. However, the intensity of this conflictuality can be, in relative values, higher or inferior to that of the activity of reference. Using data on local social and economic background enables us to test hypotheses on the "long term" relationship between the dynamics of a territory and the modes of conflictuality, and to incorporate social and human dimensions. Thus, the level of conflictuality often proves higher in areas where the levels of income and education are high, which implies that the population mobilized is well informed and educated.

Illustration of Data Crossing Using the Conflict Method. Examples from Developed and Developing Countries

The use of our methodology of analysis, and of the data contained in the *Conflict* © database led to several works[c] (some of them, marked by asterisk * are presented in the bibliography), performed on a few case studies, and to an increasing number of publications, quoted above. Our aim, here, is not to enter into all the details of these studies, but to provide examples of studies based on different areas, in quasi opposite situations. Thus, we have chosen two examples, based on two case studies that have been studied extensively by our teams, in order to provide information on the way the method and the data base have been used but also about the options provided by them in front of different situations. The two case studies (the Greater Paris region and the Chotiari reservoir case in Pakistan) are rather different in many ways:

[c] The data contained in the data base are accessible to the interested readers if they want to contact the authors. The detailed structure of the data base is also available (Galman and the participants of Conflict Program 2007)

- The Greater Paris region case study is located in a developed country, in a very densely populated area, in a peri-urban and very rich region, involving rural and urban inhabitants, conflicting for scarce resources;

- The Chotiari reservoir case study is located in developing country, in a sparse area, in a rural and very poor region, involving local users of land in a conflict against the loss of their land and natural resources.

Thus following subsections are devoted to the presentation of the dynamics of land use conflicts in these two areas, involving reflections on the use of our method in developed and developing countries, and about the possible adaptations to changing situations. We have also highlighted the different results provided by the use of various data, and their comparative utility.

Conflicts in developed Countries. An Example of Analysis of Conflictuality in the Greater Paris Region

The research conducted using our methodology in developed countries has led to studies of several sites located in rural and peri-urban areas in France[d]. They have resulted in several publications that jointly or alternatively use one or several of the aforementioned sources. They are of great importance with regards to the French case, because there is an ongoing debate on the role of land-use and proximity conflicts in rural and peri-urban areas. Namely: which types of conflicts occur in these zones? Are they related to public or private actors? Are they legitimate or are they part of the Nimby phenomenon? Can they be considered as signals of public decision failure? Do they prevent local development?

[d]The Seine Estuary, the Loire Estuary, the Regional Natural Park of Monts d'Ardèche, the Pays Voironnais in Isère, the Community of municipalities of Montrevel en Ain, the Cortenais, and Balagne areas in Corsica, the Puys mountains in Auvergne, the Réunion island area, the Greater Paris Region, the Charente catchment area, Montpellier's coastline, and Arcachon Bay.

An illustration of how we have used the data collected using our method, performed by using theConflict © database, is provided through the analysis of the levels of conflictuality in the Greater Paris Region, one of the areas surveyed (Darly 2009; Pham and Kirat, 2008; Pham et al. 2012), and, more particularly, through a comparison of two sources: data on the activities of the courts and the daily regional press. Paris is the national capital region of France, but also, and by far, the largest metropolitan area in France and can only be compared with two or three other metropolitan regions in Europe. Around the highly urbanised core composed of Paris and its suburbs, a peri-urban belt has received many residential and industrial activities that produce more or less urbanised rural landscapes. Within this peri-urban belt, the scarcity of well-located vacant spaces (well-connected to transport facilities and services centres) and the diversity of actors and interests that share the same rural environment raises many tensions and conflicts over farmland uses. But the peri-urban belt is also an area where several local development and planning initiatives dedicated to farmland protection and farming enterprises survival are currently carried out. So the level of conflict is quite high, and the local inhabitants are somewhat dissatisfied with decisions of building infrastructures of different types (highways, railroads, airport, windmills, new towns...) near the suburbs.

To start with, let us examine the geography of infrastructure-related conflicts[e] in the Greater Paris Region, the size of the pyramids indicating the number of legal proceedings. These conflicts are a good indicator of the process of peri-urbanization the region has been undergoing, a process that is met with much opposition from populations that reside in the areas concerned. Figure 2 shows that the infrastructure related conflicts are not randomly distributed in the Paris Region. On the contrary, they are concentrated in the area bordering the urban centre of Paris: One can see that the highly urbanized part of the Paris metropolis (Paris and its three bordering departments, or the "*petite couronne*" – little crown) seems little affected. The conflicts are revealing of the spatial constraints the Paris agglomeration faces in its process of expansion and in the construction of the infrastructures that are necessary to the implementation of new urban developments.

[e]They concern, for the most part, the projects of construction of road, highway, railway, river or airport infrastructures, as well as public facilities such as wastewater treatment plants, town halls, barracks, prisons, multipurpose halls...

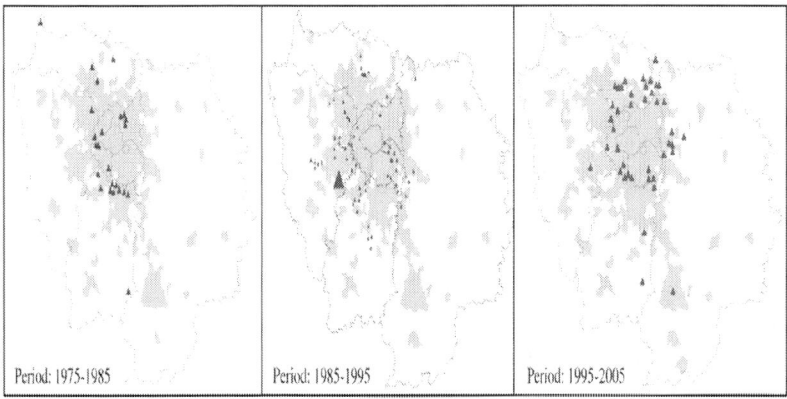

| Period: 1975-1985 | Period: 1985-1995 | Period: 1995-2005 |

Figure 2: Geography of infrastructure-related conflicts in the Greater Paris Region from 1975 to 2005 (litigation court sources).

The evolution of infrastructure-related conflicts over three successive periods - 1975–1985, 1985–1995 and 1995–2005 - corresponds to the expansion of the areas represented in grey on the map, areas which correspond to municipalities with a population of over 5000 inhabitants. Over the last thirty years, the grey area has not expanded much but that the conflicts have multiplied in different places, all located at the border of the *"petite couronne"*. They are peri-urban municipalities at the interface between the Paris agglomeration and the natural and agricultural areas which still represent over 50 percent of the total area of the Paris region. These municipalities have a high rate of urbanization (on average over 50 construction permits issued every year) and will no doubt become urban municipalities. The conflicts show that urban expansion does not always occur easily, because attempts to create infrastructures are confronted with organized opposition from residents who wish to preserve nature or their living environment. Conflicts here often take the form of opposition to decisions made by the public authorities, whether they concern the construction or the extension of infrastructure. They are a good expression of the process of legitimation of reflexivity that starts (Rosanvallon, 2008), in particular when it is based on logic of proximity with local populations exercising their democratic rights to intervene.

The *Conflicts* © database can also be used for examining the logics behind the location of conflicts such as they are revealed through the analysis of two different sources compared with the results of surveys conducted in this zone. For example, we observe heterogeneity between the DRP sources (with the daily newspaper «*Le Parisien*") and the litigation courts' data on conflicts related to infrastructures of public interest (Figure 3) or to urban development or the development of open spaces (Figure 4). The conflictual dynamics around the infrastructure identified by the local press clearly indicate that conflicts are far more intense in the western part of the Paris region where the value of land, the average income of the residents and associative activities are higher than in the rest of the region. Thus, mediatization, patent in the press articles, seems conditioned by a specific social context that gives conflicts a particular "territorial style". However, the similarities between the results obtained from both sources seem much stronger in the case of the conflicts related to urban development and the management of open spaces.

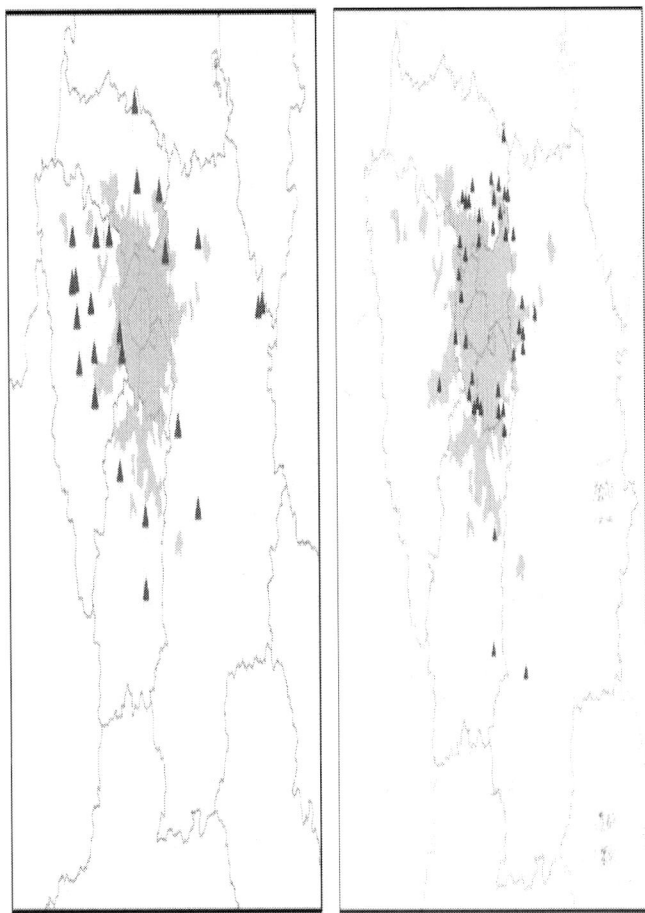

Figure 3: Conflicts related to infrastructure of public utility reported by the Daily Regional Press (2005) and litigation rulings (1999–2005) in the Greater Paris Region.

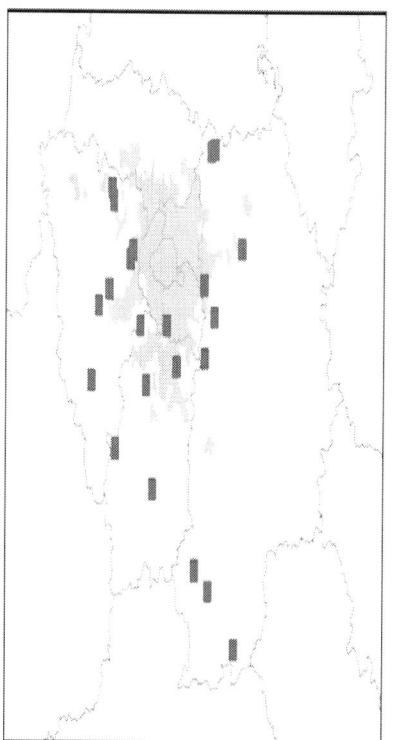

Figure 4: Conflicts related to urban development and the development of open spaces reported by the Daily Regional Press (2005) and litigation rulings (1999–2005) in the Greater Paris Region.

We can go further and undertake systematic comparisons between both sources, so as to compile complete and opposable profiles of conflictuality in the same zone. Figure 3, built from data covering the 2003–2005 period, provided by the DRP, highlights the different types of conflicts reported, as well as the number of press articles published about them. This gives us an idea of the intensity of those conflicts, as well as their respective repercussions. A distinction is made here between remedial conflicts (which start after the implementation of an infrastructure for example, or the occurrence of a contested action) and preventive conflicts (which are triggered when a project is publicized, for example, in the framework of a public enquiry pertaining to a building permit), a distinction that is possible thanks to the fact that the data have dates associated to them.

A process similar to that conducted with the DRP can be undertaken using the decisions made by tribunals, and more particularly by administrative tribunals. This leads to noticeably different results (Figure 5), which reveal the importance of using different sources of data for analyzing conflicts.

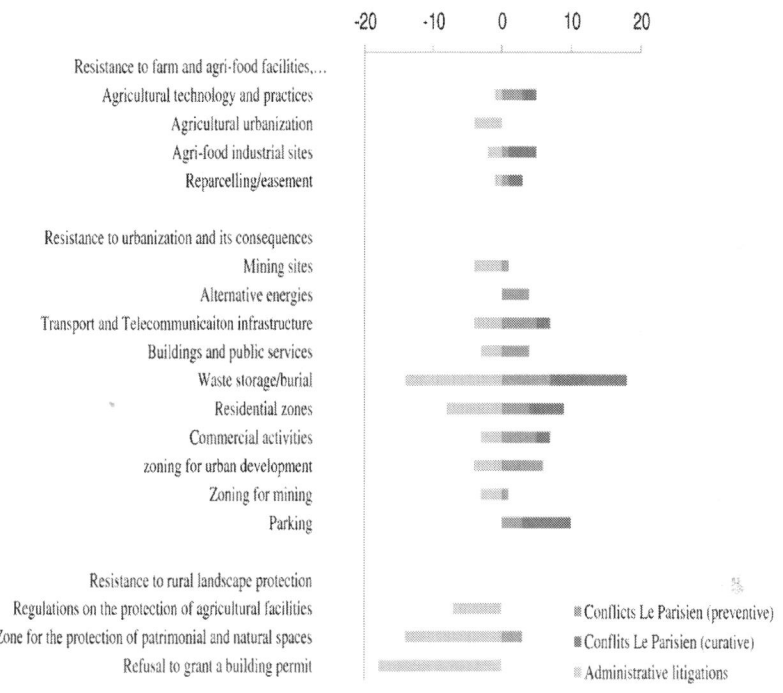

Figure 5: The main sources of land use conflicts in the Greater Paris Region, according to the administrative courts' rulings (Versailles, Cergy, Melun) (Source: *Le Parisien*, 2003–2005; Archive collections of Administrative tribunals, 2005–2006); Darly2009.

The database can also be used for analyzing more precisely certain categories of conflicts - in this case the conflicts related (directly or indirectly) to the uses of agricultural land (Darly 2009; Darly and Torre 2013a, b). We make a distinction between three main groups of conflicts according to the characteristics of the contested facilities:

the facilities dedicated to the functioning of the city – construction of transport or energy infrastructures, productive or residential activities, burial or spreading of waste, building and housing, commercial and industrial zones – the mechanisms related to agricultural economics in these territories - easements, re-parceling, and those related to landscaping projects – zones for the protection of land and natural resources, Regional Natural Parks – (Figure 6). Here again, we can see the importance of the conflicts related to the presence of the capital city (Paris) and to its expansion into peri-urban areas which regularly encroaches on land areas reserved for agricultural purposes or nature conservation and provokes the opposition of some of the people already residing in these areas.

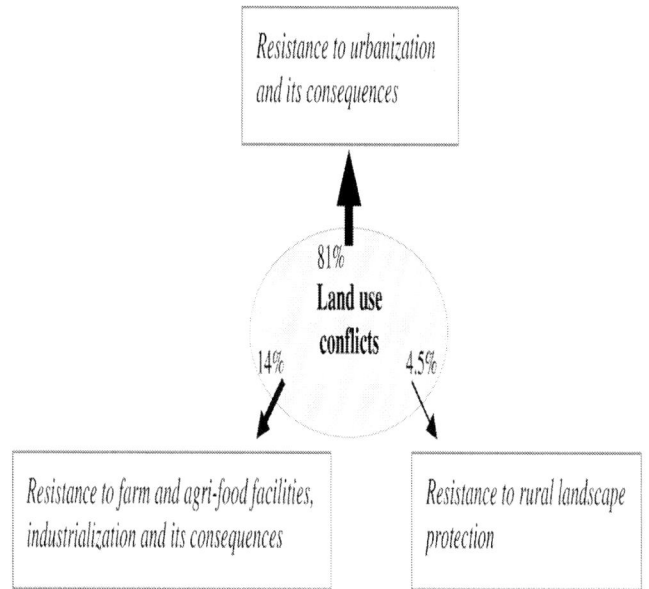

Figure 6: Conflict in the Greater Paris Region as observed by the DRP (Source: *Le Parisien*, 2003–205); Darly2009.

Comparing the results of the analysis of the DRP and of litigation ruling (Figure 7) shows noticeable differences in the case of conflicts related to the use of agricultural land: while the oppositions against the regulatory mechanisms for the protection of nature and open spaces represent a large number of litigation cases. The press reveals, more specifically, the collective and publicized dimension of the actions undertaken against so called urban sprawl interest projects (against infrastructures for example). This finding indicates that certain conflictual situations seem to lack the public dimension necessary for their publicization and therefore hardly feature in the press, but take the form of individual litigation cases. These cases mostly involve oppositions against administrative decisions such as land occupancy authorizations in urban law (building permits and planning certificates for example). Finally, a number of conflicts that have been the object of litigation concern similar categories of actors: It is the case of oppositions against some land restructuring projects, internal to the farming world. Inversely, it is mostly the confrontation between local residents and nonresident users that appears to be the main source of the conflicts that feature in the local press, conflicts in which the collective action dimension prevails.

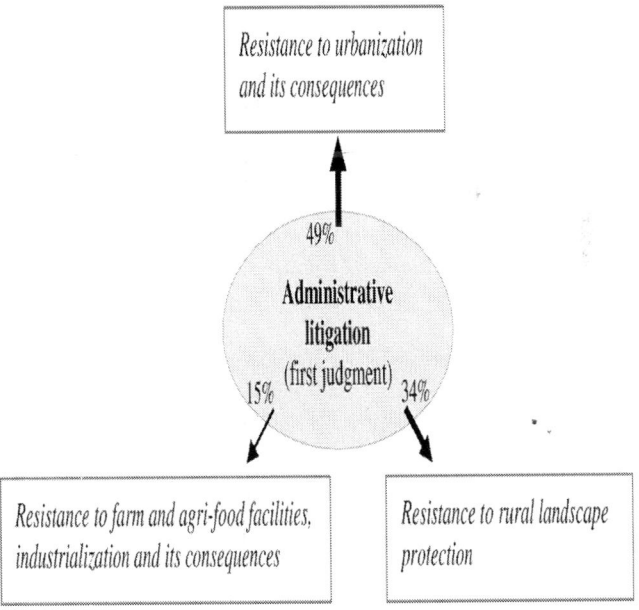

Figure 7: Conflict in the Greater Paris Region as described in legal cases dealt with by the Administrative Appeal Courts (Versailles and Paris) and the Council of State (1981–2005); Darly2009.

Conflicts in Developing Countries. An Example in South Pakistan

For a particular case study of land use conflict in developing countries, we have selected the case of Chotiari water reservoir from Pakistan; in order to put light on the land use conflicts caused by an infrastructural project setting with follow-up governance structure. This is one of Pakistan's largest infrastructure projects, which is facing opposition in the country and is held up as an example of weak governance in the planning of new infrastructures in developing countries. (Magsi and Torre, 2014) (see Figure 8).

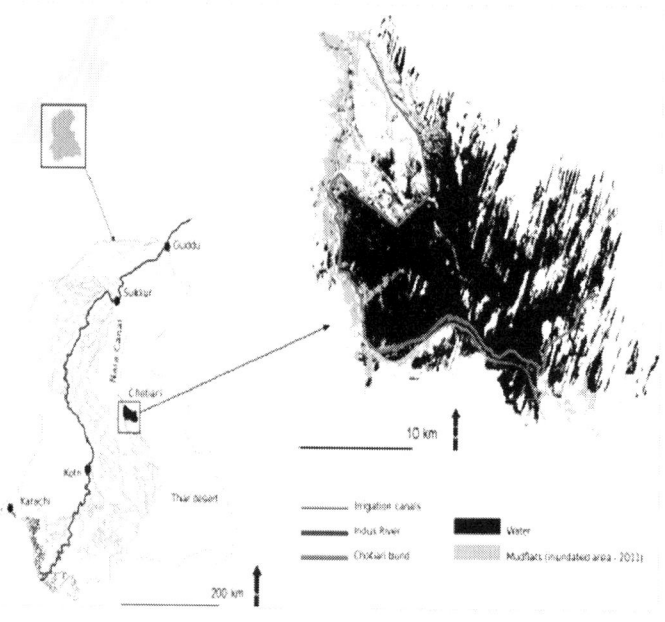

Figure 8: Location of the Chotiari water reservoir.

The Chotiari reservoir project was designed and implemented in order to increase the storage capacity of the existing lakes in the Chotiari wetlands and enable the irrigation of more arable land in Pakistan. The project was initiated in 1992 by the Water and Power Development Authority (WAPDA) and was funded by donor agencies via the World Bank. The project area extends over 18,000 hectares of entitled and unentitled land. The Chotiari reservoir area was characterized as wetlands and included lakes, forest, swamps, irrigation channels, agricultural land, barren land and a rich ecosystem, which supported the livelihoods of the local population through fishing, agriculture, grazing and a range of other economic activities. The reservoir project has created opposition between the principal actors (fishermen, farmers, livestock herders and others) on the one hand, and stakeholders from the public administration (national and provincial ministries), local politicians and landlords on the other. More specifically, a number of factors have made the task of implementing this project more complicated and controversial: the public administration's highly bureaucratic approaches and mismanagement of construction and compensation funds; local politicians' misuse of position and power with regard to forced displacements; and local landlords' exercise of power over the local population. Furthermore, opposition grew when local populations were dispossessed of their livelihoods and ancestral properties without proper compensation. In spite of all these issues, the public authorities completed and inaugurated the reservoir in February 2003, five years later than anticipated.

For this case study we have collected data through various sources, i.e. DRP, experts' opinion interviews by an open questionnaire and other sources (available literature, GIS and internet). The Chotiari project results reveal that local actors and outside stakeholders are under opposite aims and objectives of land use, and that the drivers of this situation (behaviors and interpersonal relations and actions) lead the project under situations of superposition of uses. This restlessness among stakeholders encouraged local journalists to demonstrate their issues. In fact, more than eighty percent of the articles (of total published news/articles in DRP since 1997–2011) reflected that there has been significant wrong doing associated with the land acquisitions, compensation and resettlement plans. Therefore, the project has brought on the dynamics of conflicts of land use, which has not only contributed in the changing of territoriality of the actors, but also came forth with disruption of socio-spatial practices. This interference of socio-spatial practices involved a reaction of discontent, which has sometimes expressed aggressively (Iqbal 2004). Moreover, public authorities have induced social and environmental nuisances by affecting arable lands, pasture, forest, as well as cruel displacement of local population. Besides all, the increasing water level in the reservoir is creating seepage[f], which is destroying adjacent agricultural lands (see Figure 9).

[f]The organizational management structure has allowed increasing water level in the reservoir for which purpose it has been constructed, but due to low standard earth work there is water seepage from embankments which is damaging agricultural lands outside the reservoir.

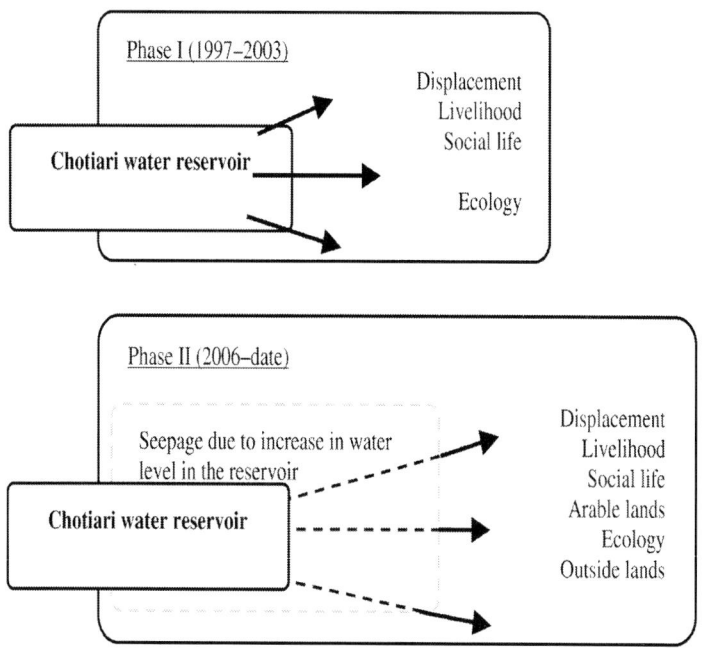

Figure 9: Conflict dynamics of Chotiari reservoir.

The multi-dimensional catastrophe of the Chotiari reservoir cannot be understood with a single factor. Therefore, it is important to visualize and quantify the structural and proximate factor dynamics with their anticipation, which have not only escalated conflicts of land use but also unrest among local population. Therefore, on the basis of articles published in daily press and opinions interviewed from experts, we came into effect to disclose the responsible factors to the conflicts of Chotiari reservoir (see Table 1). In this regard we have quantified the factors which appeared in DRP as well as in the expert opinion interviews. These factors seemed responsible for either giving favorable path to pre-conditions or conducive climate to the conflicts.

Table 1: Conflict factors of Chotiari reservoir

Factor types	Cause 0073	Percentage	
		DRP	**Experts opinion**
Structural factors	Corruption/misuse of funds	23.94	34.38
	Unilateral decision	21.81	21.88
	Lack of technical and scientific research	19.68	9.38
	International interest	7.98	12.50
	Non-existence of national resettlement policy	9.04	9.38
Proximate factors	Ethnic diversity and disarray (unrest among communities)	13.83	12.50
	Others (Nepotism, Illiteracy etc.)	3.72	0

Torre *et al.*

Torre *et al. SpringerPlus* 2014 **3**:85, doi:10.1186/2193-1801-3-85

In the above table we see the differences in the factors of conflicts highlighted by either source, which may be due to the technical approaches and scientific analysis of both sources as well as the public character of DRP data compared with the privacy of face to face interviews. Therefore, we disclose the consequences (either positive or negative) of the project after deep analyzing of DRP as well as experts' opinion in the study area (see Figure 10). After comparative analysis we found both similarities and dissimilarities in the data sources, due to privacy or the public character of their approaches.

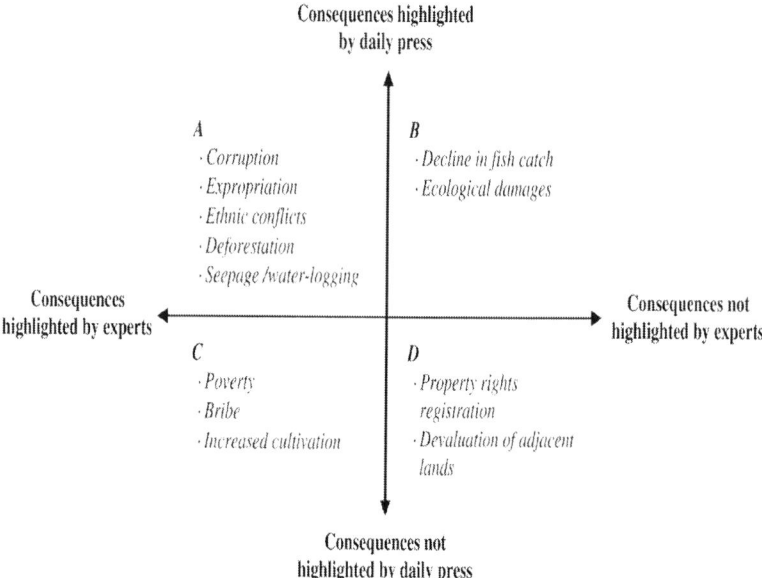

Figure 10: Consequences of Chotiari conflict: by daily press, experts and personal observations.

Results of this case study present the pattern of thought based on the expert opinions, daily press and interviews of affected households in the study area which insights a simple approach towards social representations in the categories of the actors. Thus it is difficult to categorize from planning to construction stages of the project. For example, during policy making process the actors were involved either from regional to national level, while some actors involved temporarily and did not play an active role in the administration. In order to understand the dynamic process of the project, we have tried to analyze a relational approach to provide some answers regarding social representations corresponding to a universe of interrelated elements. Though, after deep analyses we have summarized the relations, links and locations of the actors or stakeholders at different geographical scales (see Figure 11).

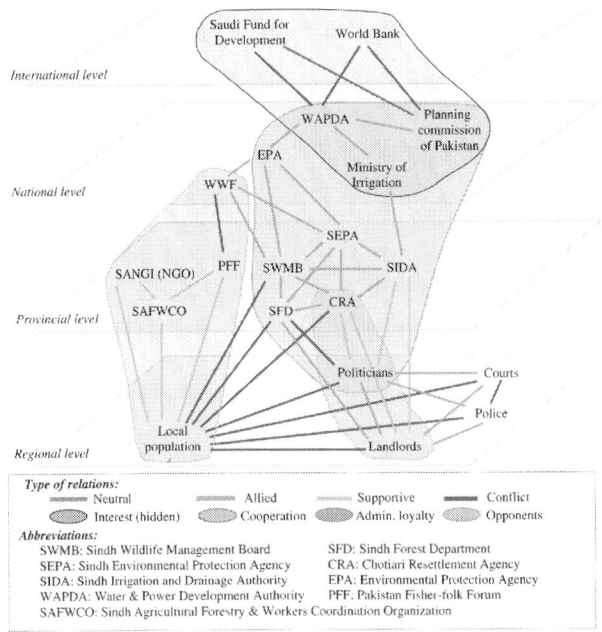

Figure 11: Networks of local actors in the Chotiari conflict.

Above figure exposes the network of the actors involved in Chotiari reservoir, with different relations and behaviors. The results show that those stakeholders have categorized themselves in defined strategies of construction and opposition of the reservoir (networks of pros and cons). Dramatically, the managing and administering actors (international to national scale) have a different representation on the reservoir area; they were found in alliance with the local politicians and landlords (regional scale), in order to construct the reservoir. According to the experts the cooperation of local stakeholders with administrative actors from international to national scales was based on some hidden interests of corruption and favoritism, while landlords and politicians have supported for construction reservoir. In this figure it is also disclosed that between two pro-construction networks (initiators and supporters), an administrative loyal network have played a key role to construct the reservoir. On the other hand, local actors were found in protestation against reservoir construction, relocation and compensation issues (Magsi and Torre, 2012). While, local population with the support of different NGOs have opposed decision of reservoir and manifested to protect natural resources of precious wetlands of Chotiari. Their limited scale of support led no valuable results of their oppositions, protests and agitations, because despite of this the reservoir has been constructed and inaugurated on February 2003. Results related to multilevel governance, we may say that this drama may be played to divert people's attention that conflicts are at regional scale only. Moreover, through above figure we have explored that law enforcement institutions (courts and police) are peripheral and seem more suppressive rather to have influence on the administration, in this situation the local population will surely have no hope of their violated rights. Through this case study we acquire the information on infrastructural development settings without agreement of local stakeholders. In fact, the only way to examine the institutional inconsistencies and distribution of dissimilar power, leading to land use conflicts and loss of local population's resources, is to analyze the dynamics of actors/stakeholders network in the study area, such as the reaction of all stakeholders during and after the project construction.

CONCLUSIONS

This article has aimed to present the work carried out in the last few years by a multidisciplinary team on the question of land use conflicts and to reveal the methodology of survey and data collection, as well as the structure of the resulting database. We have first presented the scope of our investigations by defining the conflicts in question, their characteristics, motives, manifestations and the actors involved. We have then presented our method of identification of conflicts, based on a diagnosis of the conflicts that occur in selected areas and on the combination of various methods of data collection, including interviews with experts, analyses of the DRP and of litigation rulings. Lastly, we have presented the *Conflicts* © database, with its tables and nomenclatures, which reconcile the data collected from different sources, before providing a few examples of how we have used implemented our method with cases of conflicts occurring in the Greater Paris Region.

Our studies, performed on the basis of the previous method, reveal that land use conflicts are characterized by a high diversity of expression depending on the activities, on the uses around which they emerge, on the territories in which they occur, as well as on the characteristics of the actors involved in the conflicts in question. Furthermore, some conflicts, which are closely related to certain specialized activities, are kept relatively private and may even be limited to face to face interactions between two private parties, whereas others, related to decisions concerning a large number of individuals and involving questions of land use regulations, are liable to involve the participation of the public authorities and associations representing part of the population (for example conflicts related to the definition of local urban development plans or of protected zones). The quantitative survey, crossing various sources of information, and the qualitative surveys show that land use conflicts can take extremely diverse forms of expression and manifestations, but are however centered on large categories of conflicts, territories or modes of resistance to undesirable projects. Thus, observing conflictuality has nothing to do with the mere collection of raw information expressing a reality that is easy to decrypt. The goal of the crossed analysis of various information materials is precisely to explain the fact that the modes of expression of conflicts do not just constitute «a source of information» but a framework of observation that determine the types of phenomena observed.

Our research into conflicts in rural and periurban areas shows that this dimension is key in processes of territorial management, regional development or the governance of various local activities. Sometimes, conflicts are blind oppositions or are the product of egoistical Nimby behaviours. But in many cases they constitute a way of initiating discussions on the issues and paths of territorial development and of influencing decisions by participating in processes underway from which one had been excluded. That is why they have a bearing, either on the decisions on land use and management (arbitrated negotiation) or on the composition and representativeness of the bodies responsible for taking decisions (arbitral negotiation). The conflict thus becomes an integral part of the deliberative process at the local level by allowing an expression of local democracy and the re-inclusion of participants who were forgotten or deliberately excluded during earlier project development stages.

Land-use conflicts thus constitute one form of resistance and expression of opposition to decisions that leave part of the local population unsatisfied. Some local innovations, whether technical or organizational in nature, give rise to resistance which can turn into conflict. Major changes requiring reconfiguration of the use of space (creation of transport, energy or waste-processing infrastructure, new urban master plans, territorial or environmental zoning, etc.) generate conflicts whose spatial and social extent can quickly grow. Conflicts are signals of social, technological and economic changes, indicators of novelty and innovations. They demonstrate the opposition aroused by the latter, lead to discussions on their implementations and their possible (non-) acceptability as well as on the adoption of governance procedures and their transformation under the influence of the dynamics of change. All changes encounter opposition or resistance of varying relevance and justification. But it would however be simplistic to see this resistance as a systemic sign of reactionary opposition to change because, in a number of cases, they are more a reflection of differences over the direction taken by the new initiatives that are being imposed on the public than of a stubborn desire to maintain the status quo. During these phases of conflict, social and interest groups tend to reconstitute themselves and may even undergo technical or legal changes. Once a conflict ends, it leaves behind new local agreements, new modes of governance, new configurations of discussion forums as well as new technical procedures (changes in direction, various adjustments, changes in urban planning documents, etc.), all arrived at during the negotiations.

Territorial governance processes are today undergoing intense upheavals and are subject to intense periods of discussions and conflict oppositions. These latter shape the phases of territorial innovation and thus change the directions of development and growth in rural or urban territories. Such governance mechanisms and their conflict sides can be viewed as laboratories of change because they accompany and sometimes anticipate the changes underway in the territories by giving them shape, by helping maintain a dialogue and expressions of opposition and by preventing violent confrontations or failures of development due to sluggishness or expatriation. Therefore, these changes in land use occupations and the subsequent oppositions they gave birth to are embodied in the opposing and twin forms of conflict and consultation which constitute the modes of expression and the vehicles of transmission of on-going innovations at the territorial level.

AUTHORS' CONTRIBUTIONS

AT inspired and directed the program, and wrote most of the article. RM greatly participated to the elaboration of the data collection method and to the data base elaboration, and wrote most of the jurisdictional parts of the article. HM elaborated the method for developing countries and wrote most of the developing countries example section (Chotiari, Pakistan). LB participated to the elaboration of the data collection method and to the data base elaboration, and performed several case studies. AC participated to the elaboration of the data collection method and to the data base elaboration, and performed several case studies. AC participated to the elaboration of the data collection method and to the data base elaboration, and performed several case studies. SD participated to the elaboration of the data collection method and to the data base elaboration, and wrote most of the developed countries example section (Ile-de-France, France). PhJ participated to the elaboration of the data collection method and to the data base elaboration, and performed several case studies. TK inspired and directed the program, greatly participated to the elaboration of the data collection method and to the data base elaboration, and performed several case studies. HVP participated to the elaboration of the data collection method and to the data base elaboration, and to the statistical study of litigation issues. OK participated in the elaboration of the method for developing countries and performed several case studies. All authors read and approved the final manuscript.

ACKNOWLEDGEMENTS

Some published outputs of the research program: Anthopoulou T, Moissidis A (2002) La périurbanisation dans l'espace rural grec: crise ou adaptation? Géocarrefour 7(4):359–366; Bonin M, Torre A (2004) Typologie de liens à l'espace impliqués dans les conflits d'usage, Etude de cas dans les Monts d'Ardèche. Les Cahiers de la multifonctionnalité 5:17–31; Bossuet L, Torre A (2009) Le devenir des ruralités, entre conflits et nouvelles alliances autour des patrimoines locaux. Economie Rurale 313–314:147–162;; Cadoret A (2006) Conflits d'usage liés à l'environnement et réseaux sociaux : enjeux d'une gestion intégrés ?

Le cas du littoral du Languedoc-Roussillon. PhD thesis of Geography. Université de Montpellier 3, France; Caron A, Torre A (2002) Les conflits d'usage dans les espaces ruraux. Une analyse économique. In: Perrier-Cornet P (ed) A qui appartient l'espace rural? Editions de l'Aube, Paris; Caron A, Torre A (2006) Vers une analyse des dimensions négatives de la proximité. In: Les conflits d'usage et de voisinage dans les espaces naturels et ruraux. Développement Durable et Territoires, n°7, online; Darly S (2008a) La spatialité des conflits d›usage au sein des zones périurbaines en Ile-de-France: analyse empirique d›une modalité peu connue de la gouvernance des territoires. Norois 209(4):127–146; Darly S (2008b) Tensions et conflits d'usage liés à l'agriculture. Géographie de deux corpus d'observation au sein de la région Ile-de-France. In: Loudiyi S, Bryant CR, Laurens L (ed) Territoires périurbains et Gouvernance. Perspectives de recherche, Université de Montréal, mai, pp 109–117; Darly S (2008c) Conflits d›usage et aires conflictuelles à l›échelle d›une région métropolitaine. In: Gorgeon C, Laudier I (ed) Le cas des enjeux liés à l›espace agricole en Ile-de-France. Territoires et identités en mutation, l›Harmattan, Paris, pp 87–106; Darly S, Torre A (2008) Conflits liés aux espaces agricoles et périmètres de gouvernance en Ile-de-France (résultats à partir d'analyses de la presse quotidienne régionale et d'enquêtes de terrain). Geocarrefour 83(4):307–319; Jeanneaux P, Perrier-Cornet P (2008) Les conflits d'usage du cadre de vie dans les espaces ruraux et la décision publique locale, Éléments pour une analyse économique. Economie rurale 306:39–54; Jeanneaux P, Sabau C (2009) Conflits environnementaux et décisions juridictionnelles : que nous apprend l'analyse du contentieux judiciaire dans un département français ? VertigO - la revue électronique en sciences de l›environnement 9:1. URL: http://vertigo.revues.org/index8412.html; Kirat T, Melot R (2006) Du réalisme dans l'analyse des conflits d'usage: les enseignements de l'étude du contentieux. Développement durable et territoire. Revue numérique; Kirat T, Torre A (ed) (2008) Territoires de Conflits. Analyses des mutations de l'occupation de l'espace, l'Harmattan, Paris; Kirat T, Torre A (2007) Quelques points de repères pour évaluer l›analyse des conflits dans les théories économiques, avec une emphase particulière sur la question spatiale. Géographie, Economie, Société 9(2):215–240;

Kirat T, Torre A (ed) (2006 & 2007) Conflits d'usages et dynamiques spatiales les antagonismes dans l'occupation des espaces périurbains et ruraux (I & II). Géographie, Economie, Société, 8(3) & 9(2); Lefranc C, Torre A (2004) Tensions, conflits et processus de gouvernance dans les espaces ruraux et périurbains français. In: Scarwell HJ, Franchomme M (ed) Contraintes environnementales et gouvernance des territoires. Eds de l'Aube, p 469; Magsi H (2013) Land use conflicts in developing countries, framing conflict resolution and prevention strategies to ensure economic growth and human welfare. The case of Chotiari water reservoir from Pakistan. PhD thesis of Economics. AgroParisTech, Paris, 209p; Magsi H, Torre A (2013) Approaches to understand land use conflicts in the developing countries. The Macrotheme Review 2(1):119–136; Paoli J-C, Melot R, Fiori A (2008) L'aménagement du territoire à l'épreuve de la décentralisation : conflits et concertation en Corse et Sardaigne. Pôle Sud: revue de science politique de l'Europe méditerranéenne 28(1):143–165; Torre A (2008) Conflits d'usage dans les espaces ruraux et périurbains. In: Monteventi WL, Deschenaux C, Tranda-Pittion N (ed) Campagne-ville. Le pas de deux, Presses Polytechniques et Universitaires Romandes, Lausanne; Torre A, Aznar O, Bonin M, Caron A, Chia E, Galman M, Guérin M, Jeanneaux P, Kirat T, Lefranc C, Melot R, Paoli JC, Salazar MI, Thinon P (2006) Conflits et tensions autour des usages de l'espace dans les territoires ruraux et périurbains, Le cas de six zones géographiques françaises. Revue d'Economie Régionale et Urbaine 3:415–453; Torre A, Caron A (2005a) Réflexions sur les dimensions négatives de la proximité : le cas des conflits d'usage et de voisinage. Economie et Institutions 6 & 7:183–220; Torre A, Caron A (2005b) Conflits d'usage et de voisinage dans les espaces ruraux. In: Torre A, Filippi M (ed) Proximités et Changements Socio-économiques dans les Mondes Ruraux. INRA éditions, Paris, p 322; Torre A, Lefranc C (2006) Les Conflits dans les zones rurales et périurbaines, Premières analyses de la Presse Quotidienne Régionale. Espaces et Sociétés 124–125(1–2):93–110; Torre A, Melot M, Bossuet L, Cadoret A, Caron A, Darly S, Jeanneaux P, Kirat T, Pham HV (2010) Comment évaluer et mesurer la conflictualité liée aux usages de l'espace ? Eléments de méthode et de repérage. VertigO - la revue électronique en sciences de l'environnement, Vol. 10, N° 1. http://vertigo.revues.org/9590.

REFERENCES

1. Azuela A, Herrera-Martin C (2010) Taking Land around the World: International Trends in Expropriation for Urban and Infrastructure Projects. In: Lall SV, Freire M, Yuen B, Rajack R, Helluin JJ (eds) Urban Land Markets, Improving Land Management for Successful Urbanization, Heidelberg: Springer Verlag.

2. Barré M-D, Aubusson de Cavarlay B, Zimolag M (2006) Dynamique du contentieux administratif. Analyse statistique de la demande enregistrée par les tribunaux administratifs. Ministère de la Justice, Paris: Rapport pour la Mission de recherche Droit et justice.

3. Boulding KE (1962) Conflict and Defense. New York: Harper and Row.

4. Cadene P (1990) L'usage des espaces périurbains. Une géographie régionale des conflits. Economie Rurale 118–119:235-267

5. Cadoret A (2009) Conflict dynamics in coastal zones: a perspective using the example of Languedoc-Roussillon (France. Journal of Coastal Conservation: planning and management 13:151-163

6. Campbell DJ, Gichohi H, Mwangi A, Chege L (2000) Land use conflict in Kajiado District, Kenya. Land Use Policy 17(4):337-348

7. Castro AP, Nielsen E (2001) Indigenous people and co-management: implications for conflict management. Environ Sci Policy 4(4):229-239

8. Charlier B (1999) La défense de l'environnement : entre espace et territoire, géographie des conflits environnementaux déclenchés en France depuis 1974. France: PhD thesis, Université de Pau et des Pays de l'Adour. p 750

9. Coser LA (1982) Les fonctions du conflit social. Paris: PUF.

10. Darly S (2009) Faire coexister ville et agriculture au sein des territoires périurbains. Antagonismes localisés et dynamiques régionales de la conflictualité. PhD Thesis of Ecole des Hautes Etudes en Sciences Sociales, Paris 457:p. + Annexes

11. Darly S, Torre A (2013) Conflicts over farmland uses and the

dynamics of "agri-urban" localities in the greater Paris region. Land Use Policy 33:90-99

12. Darly S, Torre A (2013) Land-use conflicts and the sharing of resources between urban and agricultural activities in the Greater Paris Region. Results based on information provided by the daily regional press. In: Noronha-Vaz T, Leeuwen VE, Nijkamp P (eds) Towns in a rural world, London: Ashgate.

13. Deininger K, Castagnini R (2006) Incidence and impact of land conflict in Uganda. J Econ Behav Organ 60:321-345

14. Diehl P (1991) Geography and War: A Review and Assessment of the Empirical Literature. Int Interact 17:11-27

15. Dziedzicki JM (2001) Gestion des conflits d'aménagement de l'espace : Quelle place pour les processus de médiation?. Thèse pour le doctorat d'aménagement de l'espace et urbanisme: Université de Tours, Tours.

16. Freund J (1983) Sociologie du Conflit. Paris: PUF.

17. Fisher R (1997) Interactive conflict resolution. Syracuse, New York: Syracuse University Press.

18. Galman M, the participants of Conflict Program (2007) Guide Base de données Conflits. UMR SAD-APT. p 64

19. Hensel PR (2001) Contentious issues and world politics: the management of territorial claims in the Americas, 1816–1992. Int Stud Q 45:81-109

20. Humphreys M (2005) Natural resources, conflict, and conflict resolution: uncovering the mechanisms. J Confl Resolution 49(4):508-537

21. Iqbal N (2004) Affectees of Tarbela and Chotiari Dams: A struggle for social justice. Addressing Existing Dams: United Nations Environmental Program. pp 69-72 Issue based workshop

22. Janelle D (1977) Structural Dimensions in the Geography of Locational Conflicts. Can Geogr 21:311-328

23. Jeong HW (1999) Conflict management and resolution. In: Kurtz L (ed) Encyclopedia of Violence, Peace and Conflict, vol 1. Academic Press, pp 389-400

24. Joerin F, Pelletier M, Trudelle C, Villeneuve P (2005) Analyse spatiale des conflits urbains, Enjeux et contextes dans la région de Québec, Cahiers de Géographie du Québec. Numéro

thématique Conflits, Proximité, Coopération 49(138):319-342

25. Kirat T, Torre A (2004) Modalités d'émergence et procédures de résolution des conflits d'usage autour de l'espace et des ressources naturelles. In: Analyse dans les espaces ruraux. vie et société: Chapter 2. Rapport de recherche CNRS, Programme Environnement.

26. Lascoumes P (1995) Les arbitrages publics des intérêts légitimes en matière d'environnement, L'exemple des lois Montagne et Littoral. Revue française de science politique 45(5):396-419

27. Leost P (1998) La stratégie contentieuse d'une association de protection de la nature en Bretagne. In: Le Louarn P (ed) Décision locale et droit de l'environnement, Etude comparée des cas breton et martiniquais. Rennes: Presses Universitaires de Rennes. pp 85-106

28. Leeuwen M (2010) Crisis or continuity? Framing land disputes and local conflict resolution in Burundi. Land Use Policy 27(3):753-762

29. Lewin K (1948) Resolving Social Conflicts. New York: Harper & Row.

30. Ley D, Mercer J (1980) Locational conflicts and the politics of consumption. Econ Geogr 56(2):89-109

31. Magsi H, Torre A (2014) Proximity analysis of inefficient practices and socio-spatial negligence: Evidence, evaluations and recommendations drawn from the construction of Chotiari reservoir in Pakistan. Land Use Policy 36:567-576

32. Magsi H, Torre A (2012) Social network legitimacy and Property right loopholes: Evidences from an infrastructural water project in Pakistan. J Infrastructure Dev 4(2):59-76

33. Mann C, Jeanneaux P (2009) Two Approaches for Understanding Land-Use Conflict to improve Rural Planning and Management. J Rural Community Dev 4(1):118-141

34. Melé P, Larrue C, Rosenberg M (2004) Conflits et territoires, Presses Universitaires François Rabelais. Tours: collection perspectives « Villes et territoires ».

35. Mc-Carthy JD, Mc-Phail C, Smith J (1996) Images of protest: dimensions of selection bias in media coverage of Washington

demonstrations, 1982–1991. Am Sociol Rev 39:101-112

36. Neslund N (1990) Dispute resolution: a matrix of mediation. J Dispute Resolution 2:217-266

37. Observatorio Permanente dos Conflitos Urbanos na Cidade de Rio de Janeiro (2010) http://www.observaconflitos.ippur.ufrj.br/novo/ajax/relatal.asp

38. Olzak S (1992) The dynamics of ethnic competition and conflicts. Stanford: Stanford University Press.

39. Owen L, Howard W, Waldron M (2000) Conflicts over farming practices in Canada: the role of interactive conflict resolution approaches. J Rural Stud 16:475-483

40. Pham HV, Kirat T (2008) Les conflits d'usage des espaces périurbains et le contentieux administratif - Le cas de la Région Ile-de-France. Revue d'Economie Rurale et Urbaine 5:671-700

41. Pham HV, Kirat TH, Torre A (2012) Les conflits d'usage dans les espaces ruraux et périurbains. Le cas des infrastructures franciliennes. Economie Rurale 332:9-30

42. Rosanvallon P (2008) La légitimité démocratique. Paris: Le Seuil.

43. Rucht D, Neidhardt F (1999) Methodological Issues in Collecting Protest Event Data: Unit of Analysis, Sources and Sampling, Coding Problems. In: Rucht D, Koopmans R, Neidhardt F (eds) Acts of Dissent: New Developments in the Study of Protest, Lanham: Rowman and Littlefield Publishers. pp 65-89

44. Simmel G (2008) Sociological Theory. New York: McGraw–Hill.

45. Starr H (2005) Territory, Proximity, and Spatiality: The Geography of International Conflict. Int Stud Rev 7:387-406

46. Stephenson GM (1981) Intergroup bargaining and negotiations. In: Turner JC, Giles H (eds) Intergroup Behaviour, Oxford: Basil Blackwell.

47. Struillou J-F (2004) L'application du droit pénal de l'urbanisme, In Etat de droit et urbanisme. Cahiers du Gridauh 11:87-126

48. Touraine A (1978) La voix et le regard. Paris: Seuil.

49. Trudelle C (2003) Au-delà des mouvements sociaux: une typologie relationnelle des conflits urbains. Cahiers de Géographie du Québec 47(131):223-242

50. Wieviorka M (2005) La violence. Paris: Hachette Littératures.

Chapter 6

Review of Underground Coal Gasification Technologies and Carbon Capture

Stuart J Self, Bale V Reddy, and Marc A Rosen

Faculty of Engineering and Applied Science, University of Ontario Institute of Technology, 2000 Simcoe Street North, Oshawa, ON, L1H 7K4, Canada

ABSTRACT

It is thought that the world coal reserve is close to 150 years, which only includes recoverable reserves using conventional techniques. Mining is the typical method of extracting coal, but it has been estimated that only 15% to 20% of the total coal resources can be recovered in this manner. If unrecoverable coal is considered in the reserves, the lifetime of this resource would be greatly extended, by perhaps a couple hundred years. Mining involves a large amount of time, resources, and personnel and contains many challenges such as drastic changes in landscapes, high machinery costs, elevated risk to personnel, and post-extraction transport. A new type of coal extraction method, known as underground coal gasification (UCG), that addresses most of the problems of coal mining is being investigated and implemented globally. UCG is a gasification process applied to *in situ* coal seams. UCG is very similar to aboveground gasification where syngas is produced through the same chemical reactions that occur in surface gasifiers. UCG has a large potential for providing a clean energy source through carbon capture and storage techniques and offers a unique option for CO_2 storage. This paper reviews key concepts and technologies of underground coal gasification, providing insights into this developing coal conversion method.

REVIEW

Introduction

The global energy supply is comprised of many different sources, including fossil fuels, uranium, and various alternative and renewable sources. Currently, over 85% of the global energy supply is derived from fossil fuels, and a high fossil fuel dependency appears likely to remain in the immediate future [1]. The amount of energy required globally is projected to increase due to growing population and industrialization [2,3]. Some feel that the total primary energy demand will double or even triple by the year 2050 relative to levels today, and as the energy demand continues to increase, future fossil fuel shortages are predicted [4,5].

Hammond [6] argues that fossil fuel depletion is a significant factor when considering sustainable energy systems for the future. Fossil fuel resources are finite and being consumed rapidly, beginning with the most economically attractive resources [7]. In the future, fossil fuel resource extraction and production rates are expected to peak and begin to decline [7]. Oil production is predicted to peak in 5 to 15 years and gas production within 40 years, with significant exhaustion of oil and gas reserves by the years 2050 and 2070, respectively [4,8]. As fossil fuel demand approaches supply levels, the cost of energy is anticipated to increase drastically, prompting research and technological developments for improved ways to convert more fossil fuel resources into useable reserves [9].

Currently, coal generates 41.5% of the world's electricity and provides 26.5% of global primary energy needs [1]. Coal has the largest reserves in the world of the fossil fuels and is abundant in many countries. It is thought that the world's recoverable coal reserve is close to 150 years at current production rates, but this only represents 15% to 20% of the entire resource [8]. Remaining global coal resources have recently been estimated to be 18 trillion tonnes [10]. This contrasts significantly with the typical figure of tens of billions of tonnes for recoverable reserves [4]. If unrecoverable coal is considered in the recoverable reserves, the lifetime of the resource could be extended by a couple hundred years. For this to be realized, new, economic extraction techniques need to be implemented.

Coal is conventionally extracted by mining, both underground and open pit. Mining operations require much time, personnel, and natural resources; typically, coal reserves lie too deep underground, or are otherwise too costly, to exploit using conventional mining methods. Conventional mining also has other issues including land subsidence, high machinery costs, hazardous work environments, coal transport requirements, localized flooding, and methane buildup in cellars of nearby homes [11].

Underground coal gasification (UCG) is a newer type of coal extraction that is being investigated and implemented around the world and that avoids most of the problems of mining coal. UCG involves the conversion of unmined coal, where coal seams are gasified, without mining, and synthetic gas (syngas) is produced for use in power generation or as chemical feedstock [12]. UCG limits the amount of underground work required by personnel, lowering risks of harm relative to conventional mining. Power generation and chemical processing plants can be built directly above a coal resource and use syngas produced through UCG, avoiding coal transport. UCG has the ability to significantly widen the resource base, where the energy contained within inaccessible coal reserves, considered uneconomical for recovery, could be recovered using UCG [11]. It has been estimated by the Underground Coal Gasification Partnership that around 4 trillion tonnes of otherwise unusable coal could be suitable for UCG [9].

UCG is appealing for expanding recoverable coal reserves, but as with the combustion of all fossil fuels, there are associated greenhouse gas emissions. Coal is the most carbon-intensive of all fossil fuels and has high associated CO_2 emissions [4]. The calorific value of fossil fuel sources varies, with typical values of 50 GJ/tonne for natural gas, 45 GJ/tonne for crude oil, and 30 GJ/tonne for coal [4]. Hence, coal has the highest CO_2 emissions per unit of thermal energy produced [13]. If coal is to become a major contributor in the future global energy supply, CO_2 capture and storage techniques would need to be incorporated in the process. UCG has good potential for CO_2 reduction. During gasification, CO_2 is produced, which can be captured from the syngas and stored for long terms. If UCG is successfully linked to such carbon capture and storage (CCS), a method will be available for exploiting the energy in previously unrecoverable coal reserves while satisfying standards for reducing CO_2 emissions.

The aim of this paper is to review key areas and technologies for underground coal gasification so as to provide insights into this developing coal conversion method.

Underground Coal Gasification

Brief UCG History

The concept of coal gasification has been recognized for more than 200 years and was first used during the late 1800 s to produce town gas fuel for heating and lighting applications [14]. Today, coal gasification is primarily used to provide fuel for advanced power plants and chemical feedstocks for use in the chemical industry [9]. Conventionally, coal is extracted from the ground through mining, processed, transported, and then gasified in a surface gasifier unit to produce syngas. Underground coal gasification is a combined extraction and conversion process.

Experimentation on UCG was first performed in 1912 by Sir William Ramsay in England [4]. The experiments demonstrated the potential of UCG, but Ramsay's work was interrupted by the First World War. After the war, further UCG research did not continue, since coal was relatively inexpensive and available through conventional mining techniques in Western Europe [4]. The former Soviet Union was the first to begin considerable research and development programs with respect to large-scale UCG systems in the 1930s [15,16]. UCG technology was first utilized within commercial operations by the former Soviet Union for heating and power generation applications, which has continued to implement these systems for over 50 years [17]. Even though UCG has the appearance of being commercially mature, the technology from the former Soviet Union has been gaining interest only recently, with a rapid increase in the number of pilot plants throughout the rest of the world over the last decade. There are many commercial projects entering pilot plant phase and undergoing study, in Australia, New Zealand, the USA, India, Pakistan, Canada, and Italy. National research programs are being reconsidered in the USA and the UK, after preliminary systems failed to reach commercial maturity. As of 2008, the number of UCG trials includes 200 in the former Soviet Union, 33 in the USA, and approximately 40 distributed between South Africa, China, Australia, Canada, New Zealand, India, Pakistan, and Europe [17,18].

UCG Concepts and Technology

UCG is similar to surface gasification [19], with syngas produced through the same chemical reactions [12]. The main difference is that surface gasification occurs in a manufactured reactor whereas the reactor for a UCG system is a natural geological formation containing unmined coal[19,20]. UCG also has similarities to *in situ* combustion processes applied in heavy-oil recovery and oil shale retorting, with such common operational parameters as roof/floor stability, seam continuity and permeability, and ground water influx [19,21].

The basic UCG concept is illustrated in Figure 1. UCG involves an arrangement of injection and production wells drilled into coal seams. The coal is ignited and compressed gasification agents are fed into the coal seam through injection wells which triggers and controls an *in situ* sub-stoichiometric combustion process, producing syngas [22]. Syngas is extracted using production wells and is processed for use [17]. Suitable gasifying agents include air, oxygen, steam/air, and steam/oxygen [23]. The main difference between using oxygen and atmospheric air is that utilization of oxygen increases the heating value of the syngas [24], but producing pure oxygen requires additional energy and resources.

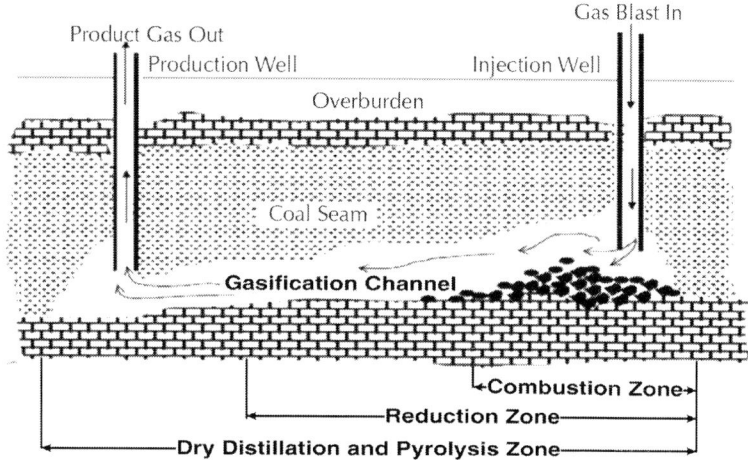

Figure 1: Schematic of *in situ* underground coal gasification process (modified from [21]).

Coal ignition is initiated through the use of an electric coil or gas firing near the face of the coal seam. Continuous oxidant flow through the injection well allows for gasification to be sustained [1]. The temperature of the gasification process is maintained through varying the oxidant flow to the reactor [23]. In UCG systems, the temperature of the coal face can reach temperatures in excess of 1,500 K [24,25].

Various chemical reactions, temperatures, pressures, and gas compositions exist at different locations within a UCG gasifier. The gasification channel is normally divided into three zones: oxidization, reduction, and dry distillation and pyrolysis [26,27]. In the oxidization zone, multiphase chemical reactions occur involving the oxygen in the gasification agents and the carbon in the coal. The highest temperatures in the gasifier occur in the oxidation zone, due to the large release of energy during the initial reactions [28]. The following reactions occur in the oxidation zone:

$$C + O_2 \rightarrow CO_2 + 393.8 \text{ kJ} \tag{1}$$

$$2C + O_2 \rightarrow 2CO + 231.4 \text{ kJ} \tag{2}$$

$$2CO + O_2 \rightarrow 2CO_2 + 571.2 \text{ kJ.} \tag{3}$$

In the reduction zone, the main reactions involve the reduction of $H_2O(g)$ and CO_2 into H_2 and CO at high temperatures within the oxidation zone [26]. The following endothermic reactions occur in the reduction zone [21,28]:

$$C + CO_2 \rightarrow 2CO - 162.4 \text{ kJ} \tag{4}$$

$$C + H_2O(g) \rightarrow CO + H_2 - 131.5 \text{ kJ.} \tag{5}$$

Under the catalytic action of coal ash and metallic oxides, a methanation reaction occurs:

$$C + 2H_2 \rightarrow CH_4 + 74.9 \text{ kJ.} \tag{6}$$

The energy terms, within the above equations, represent the amount of energy released or consumed during each reaction with the stoichiometric coefficients in equations representing moles. Equations 1 to 6 are taken from [21] and [28]. The endothermic reactions in the reduction zone decrease the temperature in the gasification channel to below that required for the reduction reactions.

Within the distillation (pyrolysis) zone, the coal seam is decomposed into multiple volatiles including H_2O, CO_2, CO, C_2H_6, CH_4, H_2, tar, and char [21,28]. At the exit of the gasification channel, the volatile composition of the syngas consists mostly of CO, H_2, and CH_4. The UCG process can also have other products, including H_2S, As, Hg, Pb, and ash [27,29,30]. The composition of syngas is highly dependent on the gasification agent, air injection method, and coal composition [31,32]. During operation, the three gasification zones move along the direction of the air flow, ensuring continuous gasification reactions [21]. A distinguishing feature of UCG, compared to surface gasification, is that drying, pyrolysis, and char gasification occur simultaneously within the coal [26].

By-products of the UCG process pose an environmental hazard to the local surroundings through leaching of organic and inorganic materials into groundwater. Environmental data were first made available after later trials in the USA, including Hanna and Hoe Creek UCG trials, for which groundwater contamination monitoring was conducted before, during, and after gasification. The results illustrated that at shallow depths, UCG can pose a significant risk to groundwater in adjacent strata [30].

Groundwater pollution around UCG zones is mainly caused by one of the following mechanisms: dispersion and penetration of the pyrolysis products of the coal seam to the surrounding rock layers, the emission and dispersion of high contaminants with gas products after gasification, and migration of residue by leaching and penetration of groundwater [30]. In addition, the escaped gases such as carbon dioxide, ammonia, and sulfide may change the pH value of the local strata if dissolved.

The entire process is confined to the space of the coal seam and is sealed from the surface by natural geological formations or man-made barriers; the coal seam and strata serve, to some extent, as a natural groundwater cleaning system. In general, systems have active pressure control, in which the cavity pressure is held in equilibrium or below that of the surrounding strata [17,30]. The pressure difference induces flow into the reactor space, which inhibits gasification products from leaking away from the cavity [33,34].

The quality of the product gas is influenced by several parameters - such as the pressure inside the coal seam, coal properties, feed conditions, kinetics, and heat and mass transport within the coal seam - and the product of the UCG process is a multi-compound, high-temperature, and high-pressure syngas [1]. When the syngas reaches the surface, it is cleaned and undesired by-products are removed from the product stream [24]. Removal techniques are similar to those used with surface gasifiers. Once the by-products are removed, they can be disposed of safely, or used for other chemical processes [16]. The degree of cleaning required is dependent on the use of the syngas; syngas is cleaned either to meet the specification for input into a gas turbine (for electricity generation) or to be of sufficient purity for use as a chemical feedstock for conversion to synthetic fuels [20].

Methods for UCG. Two standard methods of preparing a coal seam for gasification have been utilized successfully: shaft and shaftless. The method implemented is dependent on parameters such as the natural permeability of the coal seam, geochemistry of the coal, seam thickness, depth, width and inclination, proximity of urban developments, and the amount of mining desired [34].

- *Shaft UCG methods.* Shaft methods use coal mine galleries and shafts to transport gasification reagents and products, which sometimes entail the creation of shafts and the drilling of large-diameter openings through underground labor [34]. The shaft method was the first technique utilized within UCG systems. Currently, the shaft method is only employed in closed coal mines due to economic and safety reasons [34]. The following are examples of common UCG shaft methods:

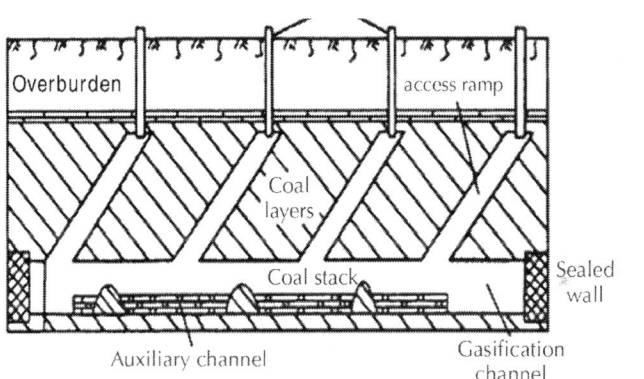

Figure 2: Schematic of the structure of a LLT underground gasifier (modified from [28]).

- *Chamber or warehouse method.* This method utilizes constructed underground galleries with brick walls separating coal panels. Gasification agents are supplied to a previously ignited coal face on one side of the wall, and the syngas is removed from a gallery on the other side. The chamber method strongly relies on the natural permeability of the coal seam to allow for sufficient oxidant flow through the system. The syngas composition may vary during operation, and the gas production rates are often low. To improve system output, coal seams are often outfitted with explosives for rubblization prior to the reaction zone [21].

- *Borehole producer method.* For this method, parallel underground galleries are created within a coal seam with sufficient distance between them. The galleries are connected by drilling boreholes from one gallery to the other [34]. Remote electric ignition of the coal in each borehole is used to initiate the gasification process. This method is designed to gasify considerably flat-lying seams. Some variations exist where linking of the galleries is accomplished through hydraulic and electric linking [21,34].

- *Stream method.* This method is designed for sharply inclined coal beds. Parallel pitched galleries following the contour of the coal seam are constructed and are connected at the bottom of the seam by a horizontal gallery also known as a fire-drift. To initiate gasification, fire is introduced within the horizontal gallery. The hot coal face moves up the seam slope with oxidant fed through one inclined gallery and syngas leaving through the other. The main advantage of this method is that the ash and roof material drop down to fill the void space created during the process, which prevents suffocating the gasification process at the coal front [21].

- *LLT gasification method.* This method utilizes mined tunnels or constructed roadways to connect the injection well to the production well [4]. Typical long and large tunnel (LLT) systems consist of a gasification channel, two auxiliary holes, and two auxiliary tunnels (Figure 2). The auxiliary holes are arranged between the injection and production wells and are used as malfunction holes for the injection of air and water vapor, or to discharge gas for added gasifier control. LLT also includes an auxiliary tunnel constructed of bricks, which is an auxiliary installation for air injection that prevents blockage in the gasification channel. The mined tunnels are isolated by sealing walls to prevent leakage of combustible gases from the gasifier [35]. The location and height of the oxidant injection points and gas outlet points can be adjusted, allowing for two-dimensional control of oxidant injection and gas production [28].

- *UCG methods.* Recently, most of the focus of research has been on the shaftless methods, which employ directional drilling techniques [6]. The preparation of a reactor for the directional drilling technique consists of the creation of dedicated in-seam boreholes for oxidant injection and product collection using drilling and completion technology that has been adapted from oil and gas production.

With shaftless methods, all preparation and operational processes are carried out through a series of boreholes drilled from the surface into a coal seam and do not require underground labor. Preparation of a shaftless reactor consists of the creation of dedicated in-seam boreholes for oxidant injection and product collection using drilling and completion technology that has been adapted from oil and gas production [34]; the approach generally includes drilling inlet and outlet boreholes into a coal seam, increasing the coal permeability between the inlet and outlet boreholes, igniting the coal seam, introducing an oxidant for gasification, and extraction of the product gas from the outlet well [21]. Currently, there are two main classifications of shaftless UCG methods: linked vertical well (LVW) and controlled retractable injection point (CRIP).

- *LVW method.* The LVW method is one of the oldest methods for UCG and is derived from technology developed in the former Soviet Union [16]. Vertical wells are drilled into a coal seam, and internal pathways in the coal are utilized to direct the oxidant and product gas flow from the inlet to the outlet borehole. Internal pathways can be naturally occurring or constructed [35]. In its simplest form, the LVW method has inlet and outlet borehole locations that are static for the life of the system. During operation, the coal face migrates and it is found that system control, performance, and syngas quality are affected negatively as the distance from the coal face to the oxidant injection point increases [4]; this factor greatly reduces the feasibility of simple LVW systems.

A more advanced LVW approach involves a series of dedicated injection boreholes located along the length of a coal seam [21]. Over the life of a UCG reactor, the coal face, being gasified, travels as localized coal is exhausted [4]. Having multiple boreholes for injection allows for improved static operating conditions. A more complex variation of the LVW method also exists where multiple inlet and outlet boreholes are drilled into a coal seam, forming inlet and outlet borehole pairs. Parallel inlet and outlet manifolds are connected to the boreholes to provide a path for oxidant and syngas flows, respectively. Coal between each pair of inlet and outlet boreholes forms a zone. When the coal in a zone has been exhausted, new boreholes are drilled in a location of fresh coal, forming new zones[21].

Low-rank coals, such as lignites, have considerable natural permeability and can be exploited for UCG without the need for linking technologies. However, high-rank coals, such as anthracites, are far less permeable, making the gas production rate more limited if UCG is employed [35]. For the use of high-rank coals in UCG, a method of linking must be employed to increase the permeability and fracture the coal seam [36]. The boreholes in traditional LVW gasifiers are linked by special methods including forward combustion, reverse combustion, fire linkage, electric linkage, hydrofracturing, and directional drilling to create sizable gasification channels [35,37].

Controlled retractable injection point. Over the span of a coal seam, the geometry may change, resulting in variable UCG operation and system performance [38]. In the past, this problem was solved by having multiple injection and/or production wells so that static operating conditions could be accomplished through moving the gasifier zones to fresh coal [16]. CRIP offers an alternative approach where the vertical injection well is not moved, but the injection point is moved within the coal seam to fresh coal when necessary [39].

The CRIP method relies on a combination of conventional drilling and directional drilling to access the coal seam and physically form a link between the injection and production wells, without the use of linking technologies utilized in LVW methods [38]. A vertical section of injection well is drilled to a predetermined depth, after which directional drilling is used to expand the hole and drill along the bottom of the coal seam creating a horizontal injection well [40]. At the end of the injection well, a gasification cavity is initiated in a horizontal section of the coal seam, creating a localized reactor. The CRIP system utilizes a burner attached to retractable coiled tubing which is used to ignite the coal[39]. The burner burns through the borehole casing to ignite the coal. The ignition point can be moved to any desired location along the horizontal injection well for the creation of a new gasification cavity after a deteriorating reactor has been deserted [38]. Typically, the injection point is retracted using a gas burner, which burns a section of the liner at a desired location [39]. In this manner, accurate control of the gasification process can be obtained. This UCG method has gained popularity in Europe and the USA, but the use of the CRIP method for UCG is fairly new and currently has not become commonly employed [4].

UCG with CO_2 Capture and Storage

All fossil fuels emit CO_2 when combusted. Currently, coal has the highest CO_2 emissions, per unit energy produced, of the fossil fuels used in combustion [13,41]. To maintain and expand the use of coal, implementation of CCS technologies is becoming imperative.

CO_2 capture can be performed in three main fashions: pre-combustion, post-combustion, and oxy-firing [42]. A broad range of technology options are available for capturing CO_2 including physical absorption, chemical absorption, membrane separation, and cryogenic separation [42,43]. Within UCG, the syngas compositions, temperatures, and pressures of production streams at the exit of a production well are comparable to those of surface gasifiers, which allow similar methods of CO_2 capture. Due to similarities, it is believed that UCG syngas could take advantage of separation using physical sorbents, within a pre-combustion arrangement, which has costs comparable to capture technologies commonly utilized in integrated gasification combined cycles [4,12]. Post-combustion methods are also applicable and would be directly comparable in terms of cost and performance to typical post-combustion systems utilized in power plants. Oxy-firing options are possible for UCG as well, and within a power generating scenario, an air separation unit can generate O_2 streams for injection into the UCG and for use in an oxy-fired plant utilizing the syngas [12].

The spatial coincidence of geological carbon storage (GCS) options with UCG opportunities suggests that designers could colocate and combine UCG and GCS systems with high potential for effective CO_2 storage [18]. In general, these storage options would be the same for conventional carbon sequestration operations, including saline formations and mature oil and gas fields [44]. For UCG-CCS utilizing conventional sequestration options, there could exist common interests in site characterization and monitoring between UCG and CCS projects, where work performed during the design and implementation of one project could be used within the other. Coordinating UCG and CCS designs would improve economics for both projects.

If UCG and CCS are coupled, there is an attractive carbon management scheme associated, where most of the expected CO_2 emissions are sequestered back into a coal seam void that has been recently created by spent subsurface reactors through existing injection and production wells[13,21]. When voids are created, they typically collapse, similar to voids produced during longwall coal mining, leaving zones of artificial breccias with high permeability. Suitable containment zones prevent vertical flow of CO_2 to the surface, where storage locations are isolated from the surface by low-permeability strata (known as seals or caprocks, often shales or evaporites) [4,45]. For a spent UCG system to accommodate CO_2 storage, the void must be at depths below approximately 800 to 1,000 m [44-46]. These depths are required so that supercritical pressures and temperatures exist that allow the CO_2 density to be high enough (approximately 500 to 700 kg/m^3) to limit the storage volume required [45].

The UCG-CCS approach, if successful, could offer an integrated energy recovery and CO_2 storage system, which exploits a new sequestration resource created during operation. A significant challenge with CCS is its large energy requirement [47], of which a considerable portion is consumed during CO_2 capture and compression [48]. The pressure after compression is generally high enough to allow for a reduction in pressure during transport while allowing the fluid to be in a liquid state [9]. If CO_2 storage is accommodated in spent UCG reactors, CO_2 transport and compression requirements decline. CO_2 transport accounts for 5% to 15% of a conventional CCS financial budget, which can be lowered with a self-contained UCG-CCS project, through reduced piping and shipping requirements associated with long-distance transport [18]. A large portion of the budget for a CCS project is allotted for CO_2 storage, typically 10% to 30%, most of which is used for geological and geophysical studies and drilling injection wells [18,48]. These tasks are commonly completed during UCG construction and would not need to be repeated for the implementation of CCS, thus reducing system cost relative to conventional storage methods [18].

As of 2009, it remains unclear if CCS using UCG-produced voids is viable [44]. Until recently, this alternative has received little attention, and there remains substantial scientific uncertainty associated with the technological challenges and environmental risks of storing CO_2 in this manner [13,44]. For full-scale commercialization, extensive research and development is needed to alleviate the uncertainties. Currently, CO_2 sequestration is under development internationally by such organizations as the Intergovernmental Panel on Climate Change and Carbon Sequestration Leadership Forum [13].

CONCLUSIONS

Although the earth is an abundant source of coal, a significant amount is currently unrecoverable. With the introduction of UCG, recoverable coal reserves can be expanded by possibly a couple hundred years. Coal is likely to remain used in many countries, increasing the needs for new technologies that permit more environmentally benign extraction and utilization. Wide-scale use of UCG is such a technology option, with the syngas it produces usable as a fuel. Fossil fuels typically utilized in power production could then be used for other purposes, which would result in large reductions in their consumption rates.

UCG offers a coal extraction and conversion method in a single process that avoids many of the challenges associated with conventional mining practices. UCG has a high potential for integration with CCS using conventional methods utilized in power production due to similarities with surface gasifier units. UCG also has the potential to store CO_2 within voids created during its operation, which reduces the need for transport and storage site identification. In essence, UCG could provide a cost-effective, near-zero-carbon, energy source through the use of a self-contained system with a closed carbon loop.

AUTHORS' CONTRIBUTIONS

SS carried out the review and drafted the manuscript. BR and MR co-supervised the investigation, conceived of the study, reviewed the manuscript, and helped draft parts. All authors read and approved the final manuscript.

ACKNOWLEDGMENTS

The authors acknowledge the financial support of the Natural Sciences and Engineering Research Council of Canada.

REFERENCES

1. Daggupati, S, Mandapati, RN, Mahajani, SM, Ganesh, A, Pal, AK, Sharma, RK, Aghalayam, P: Compartment modeling for flow characterization of underground coal gasification cavity. Ind. Eng. Chem. Res.. 50, 277–290 (2011).

2. Tanaka, N(e): World Energy Outlook 2009, International Energy Agency, Paris (2009)

3. Energy Information Administration: International Energy Outlook 2010, Department of Energy, U.S. Government, Washington (2010)

4. Roddy, DJ, Younger, PL: Underground coal gasification with CCS: a pathway to decarbonising industry. Energy Environ. Sci.. 3, 400–407 (2010).

5. Ediger, VS, Hosgor, E, Surmeli, AN, Tatlidil, H: Fossil fuel sustainability index: an application of resource management. Energy Policy. 35, 2969–2977 (2007).

6. Hammond, GP: Energy, environment and sustainable development: a UK perspective. Transactions of the Institution of Chemical Engineers, Part B. 78, 304–323 (2000)

7. Aleklett, K, Hook, M, Jakobsson, K, Lardelli, M, Snowden, S, Soderbergh, B: The peak of the oil age–analyzing the world oil production reference scenario in World Energy Outlook 2008. Energy Policy. 38, 1398–1414 (2010).

8. World Energy, C: Survey of Energy Resources 2007, London, World Energy Council (2007)

9. Ghose, MK, Paul, B: Underground coal gasification: a neglected option. Int. J. Environ. Stud..64(6), 777–783 (2007).

10. Couch, G: Underground Coal Gasification, IEA Clean Coal Centre, London (2009)

11. Shackley, S, Mander, S, Reiche, A: Public perceptions of underground coal gasification in the United Kingdom. Energy Policy. 34, 3423–3433 (2006).

12. Burton, E, Friedmann, J, Upadhye, R: Best Practices in Underground Coal Gasification, Lawrence Livermore National Laboratory, Livermore (2006)

13. Khadse, A, Qayyumi, M, Mahajani, S, Aghalayam, P: Underground coal gasification: a new clean coal utilization technique for India. Energy. 32, 2061–2071 (2007).

14. Breault, RW: Gasification processes old and new: a basic review of the major technologies. Energies. 3, 216–240 (2010).

15. Gregg, DW, Hill, RW, Olness, DU: An Overview of the Soviet Effort in Underground Gasification of Coal, Lawrence Livermore National Laboratory, Livermore (1976)

16. Shafirovich, E, Varma, A: Underground coal gasification: a brief review of current status. Ind. Eng. Chem. Res.. 48, 7865–7875 (2009).

17. Van der Riet, M: Underground coal gasification. Proceedings of the SAIEE Generation Conference, Eskom College, Midrand (19 Feb 2008)

18. Roddy, D, Gonzalez, G: Underground coal gasification (UCG) with carbon capture and storage (CCS). In: Hester RE, Harrison RM (eds.) Issues in Environmental Science and Technology, pp. 102–125. Royal Society of Chemistry, Cambridge (2010)

19. Pana, C: Review of Underground Coal Gasification with Reference to Alberta's Potential, Energy Resources Conservation Board, Edmonton (2009)

20. Walker, L: Underground coal gasification: a clean coal technology ready for development. The Australian Coal Review. 8, 19–21 (1999)

21. Lee, S, Speight, JG, Loyalka, SK: Handbook of Alternative Fuel Technologies, CRC, Boca Raton (2007)

22. Kempka, T, Plötz, ML, Schlüter, R, Hamann, J, Deowan, SA, Azzam, R: Carbon dioxide utilisation for carbamide production by application of the coupled UCG-urea process. Energy Procedia. 4, 2200–2205 (2011)

23. Perkins, G, Sahajwallaa, V: Modelling of heat and mass transport phenomena and chemical reaction in underground coal gasification. Chem. Eng. Res. Des.. 85(3), 329–343 (2007).

24. Perkins, G, Sahajwalla, V: Steady-state model for estimating gas production from underground coal gasification. Energy Fuel. 22, 3902–3914 (2008).

25. Peng, FF, Lee, IC, Yang, RYK: Reactivities of in situ and ex situ coal chars during gasification in steam at 1000-1400 °C. Fuel Process. Technol.. 41, 233–251 (1995).

26. Perkins, G, Sahajwallaa, V: A mathematical model for the chemical reaction of a semi-infinite block of coal in underground coal gasification. Energy Fuel. 19, 1679–1692 (2005).

27. Yang, LH, Pang, XL, Liu, SQ, Chen, F: Temperature and gas pressure features in the temperature-control blasting underground coal gasification. Energy Sources: Part A. 32, 1737–1746 (2010).

28. Yang, L, Liang, J, Yu, L: Clean coal technology—study on the pilot project experiment of underground coal gasification. Energy. 28, 1445–1460 (2003).

29. Liu, S, Wang, Y, Yu, L, Oakey, J: Thermodynamic equilibrium study of trace element transformation during underground coal gasification. Fuel Process. Technol.. 87, 209–215 (2006).

30. Shu-qin, L, Jing-gang, L, Mei, M, Dong-lin, D: Groundwater pollution from underground coal gasification. J. China Univ. Mining & Technol.. 17(4), 0467–0472 (2007).

31. Sta czyk, K, Howaniec, N, Smoli ski, A, wiadrowski, J, Kapusta, K, Wiatowski, M, Grabowski, J, Rogut, J: Gasification of lignite and hard coal with air and oxygen enriched air in a pilot scale ex situ reactor for underground gasification. Fuel. 90, 1953–1962 (2011).

32. Prabu, V, Jayanti, S: Integration of underground coal gasification with a solid oxide fuel cell system for clean coal utilization. Int. Journal of Hydrogen Energy. 37, 1677–1688 (2012).

33. Yang, LH: A review of the factors influencing the physicochemical characteristics of underground coal gasification. Energy Sources, Part A. 30(11), 1038–1049 (2008).

34. Wiatowski, M, Sta czyk, K, wi drowski, J, Kapusta, K, Cybulski, K, Krause, E, Grabowski, J, Rogut, J, Howaniec, N, Smoli ski, A: Semi-technical underground coal gasification (UCG) using the shaft method in Experimental Mine "Barbara". Fuel. 99, 170–179 (2012)

35. Liang, J, Liu, S, Yu, L: Trial study on underground coal gasification of abandoned coal resource. In: Xie H, Golosinki TS (eds.) Proceedings of the ‹99 International Symposium on Mining

Science and Technology, Beijing, August 1999. Mining Science and Technology 99, pp. 271–275. A.A. Balkema, Rotterdam (1999).

36. Blinderman, MS, Klimenko, AY: Theory of reverse combustion linking. Combustion and Flame.150, 232–245 (2007).

37. Blinderman, MS, Saulov, DN, Klimenko, AY: Forward and reverse combustion linking in underground coal gasification. Energy. 33, 446–454 (2008).

38. Nourozieh, H, Kariznovi, M, Chen, Z, Abedi, J: Simulation study of underground coal gasification in Alberta reservoirs: geological structure and process modeling. Energy Fuel. 24, 3540–3550 (2010).

39. Klimenko, AY: Early ideas in underground coal gasification and their evolution. Energies. 2, 456–476 (2009).

40. Wang, GX, Wang, ZT, Feng, B, Rudolph, V, Jiao, JL: Semi-industrial tests on enhanced underground coal gasification at Zhong-Liang-Shan coal mine. Asia-Pac. J. Chem. Eng.. 4, 771–779 (2009).

41. Nag, B, Parikh, J: Indicators of carbon emission intensity from commercial energy use in India. Energy Economics. 22, 441–461 (2000).

42. Göttlicher, G, Pruschek, R: Comparison of CO2 removal systems for fossil-fuelled power plant processes. Energy Convers. Mgmt.. 38, 173–178 (1997).

43. Ho, MT, Allinson, G, Wiley, DE: Comparison of CO_2 separation options for geo-sequestration: are membranes competitive?. Desalination. 192, 288–295 (2006).

44. Friedmann, SJ, Upadhye, R, Kong, FM: Prospects for underground coal gasification in carbon-constrained world. Energy Procedia. 1(1), 4551–4557 (2009).

45. Orr, FM: Onshore geologic storage of CO_2. Science. 325(5948), 1656–1658 (2009

46. Budzianowski, WM: Value-added carbon management technologies for low CO2 intensive carbon-based energy vectors. Energy. 41, 280–297 (2012).

47. Steinberg, M: Fossil fuel decarbonization technology for mitigating global warming. International Journal of Hydrogen Energy. 24, 771–777 (1999).

48. Gibbins, J, Chalmers, H: Carbon capture and storage. Energy Policy. 36, 4317–4322 (2008).

Optimization of the Self-Propagating High-Temperature Process for the Fabrication in Situ of Lunar Construction Materials

Gianluca Corrias, Roberta Licheri, Roberto Orrù, and Giacomo Cao

Dipartimento di Ingegneria Meccanica, Chimica e dei Materiali, Centro Studi sulle Reazioni Autopropaganti (CESRA), Unità di Ricerca del Consorzio Interuniversitario Nazionale per la Scienza e Tecnologia dei Materiali (INSTM), Unità di Ricerca del Consiglio Nazionale delle Ricerche (CNR), Dipartimento di Energia e Trasporti - Università degli Studi di Cagliari, Piazza D'Armi, 09123 Cagliari, Italy

ABSTRACT

The highly exothermic self-propagating thermite reduction of ilmenite ($FeTiO_3$) is systematically investigated for the *in situ* fabrication of composite ceramics in Lunar environment. Because of its relatively lower volatility, Al is preferred to Mg as reducing agent. A self-propagating (SHS) behavior is displayed only if the $Al/FeTiO_3$ molar ratio is higher than 0.9. In addition, when the amount of the reducing metal in the mixture is increased, the reactive process proceeds faster and the combustion temperature becomes higher, as a consequence of the increased system exothermicity. Correspondingly, the maximum amount of Lunar regolith (containing up to 20 wt.% ilmenite) to be possibly reacted with additional $FeTiO_3$ and Al is identified. The obtained product, consisting of a complex mixture of various Al-, Ti-, Mg-, and Ca-oxides along with metallic and intermetallic phases, displays good compressive strength properties (25.8–27.2 MPa) that make it promising as construction material. Parabolic flight experiments evidenced that neither SHS process dynamics nor product characteristics are significantly affected by gravity. All the obtained findings allows us to conclude that the optimal conditions identified during terrestrial experiments are also valid for *in situ* applications in Lunar environment, with the only exception for the slight overpressure (few Torr) required for limiting Al vaporization during SHS.

INTRODUCTION

It is well recognized that human exploration on the Moon, Mars and near Earth asteroids, etc., could be strongly facilitated and significantly time extended by the possibility of "In Situ Fabrication and Repair" (ISRU) infrastructures and equipments for satisfying the needs encountered during future Space Missions[1], [2] and [3]. Moreover, to overcome the obvious difficulties and costs related to the transportation of the required material from the Earth, the so-called "In Situ Fabrication and Repair" (ISFR) paradigm is recommended in combination with the ISRU concept.

In this context, several technologies that make use of available *in situ* Lunar resources have been proposed in the last two decades by groups of scientists and engineers, with the aim of developing suitable constructions to be placed on the Moon for the protection against cosmic rays, solar wind, meteoroids, etc.[4], [5], [6], [7], [8], [9], [10] and [11].

Most of the proposed methods [7], [8], [9], [10] and [11] are based on the occurrence of combustion synthesis-type reactions for the fabrication of ceramic-based products using Lunar regolith. For instance, the preparation of Lunar bricks was investigated starting from a mixture of JSC-1 Lunar regolith simulant, Ti and B, according to the exothermic chemical reaction $x(Ti + 2B) + (1 - x)JSC-1 \rightarrow x(TiB_2) + (1 - x)JSC-1$, that displays a self-propagating high temperature synthesis (SHS) behavior when $x > 0.25$ [7]. Another recently proposed route is represented by the direct aluminothermic reduction of Lunar regolith simulant for producing ceramic composite materials [8] and [9]. Apparently, this system does not behave like in a classical SHS process, where a local and very rapid ignition is able to generate a self-propagating combustion front [12] and [13]. In contrast, a relatively long preheating step (7–15 min) was needed in the previously cited studies [8] and [9] to activate the thermite reaction in the starting mixture.

On the other hand, as recently reported in the literature [10] and [14], the occurrence of self-sustaining combustion synthesis reactions is possible in mixtures obtained after blending JSC regolith with suitable amounts of ilmenite ($FeTiO_3$), to simulate the enrichment of Lunar soils in this species, and Al, as reducing agent. This process takes advantage of the strongly exothermic character of the aluminothermic reduction of ilmenite ($FeTiO_3$), whose presence is up to 20 wt.% on Moon soil [15]. Thus, the use of available *in situ* resources represents an important aspect in the framework of the ISRU principle. In this context, it is important to note that the reduction of $FeTiO_3$ by SHS was addressed in the literature according to the following reactions: $FeTiO_3 + 3\,Mg \rightarrow 3MgO + TiFe$ [16] and $FeTiO_3 + 7Al + 3C \rightarrow 3Al_2O_3 + TiC + Fe_3Al$ [17].

In a recent investigation [11], where a similar approach was adopted, the use of magnesium as reducing agent was considered preferable, in comparison with aluminum, on the basis of the higher adiabatic temperature of the related system.

The versatility of the SHS technology for ISRU and ISFR applications was emphasized in a paper review [18], where the possible exploitation of this synthesis method for the fabrication of structural components, acoustic damping and in-space sterilization of materials coming from Mars was specifically addressed.

The self-propagating metallo-thermic reduction of Lunar regolith enriched in iron-titanate is taken into account in the present work. Specifically, the choice of the reducing agent and the influence of the most important processing parameters (composition of the starting mixture, gas pressure level, and gravity conditions) are examined in a systematic manner. The motivation for this study is the optimization of the SHS process for evaluating its possible utilization under the atmospheric and gravity conditions encountered on the Moon, that are quite different from the Earth (cf. Table 1).

Table 1: Comparison between typical conditions on Moon and Earth [19]

	Moon	Earth
Gravitational acceleration (m s−2)	1.622	9.81
Temperature range at the equator (°C)	−173 to 127	0 to 60
Temperature range at poles (°C)	−258 to 113	−89.2 to −18
Atmospheric pressure at surface (Torr)	$1 \times 10-12$	760

EXPERIMENTAL MATERIALS AND METHODS

General Issues

The JSC Lunar regolith simulant utilized in this work was provided by Orbital Technologies Corporation (Madison, WI, USA).

The XRD analysis of the as received simulant is reported in Fig. 1. Plagioclase, specifically anorthite ($CaAl_2Si_2O_8$), Ca-rich pyroxene, in particular wollastonite ($CaSiO_3$), and olivine, namely fosterite (Mg_2SiO_4), are the main phases detected in the resulting, rather complex, pattern. Ilmenite minerals and other minor phases are also generally present among regolith constituents [15].

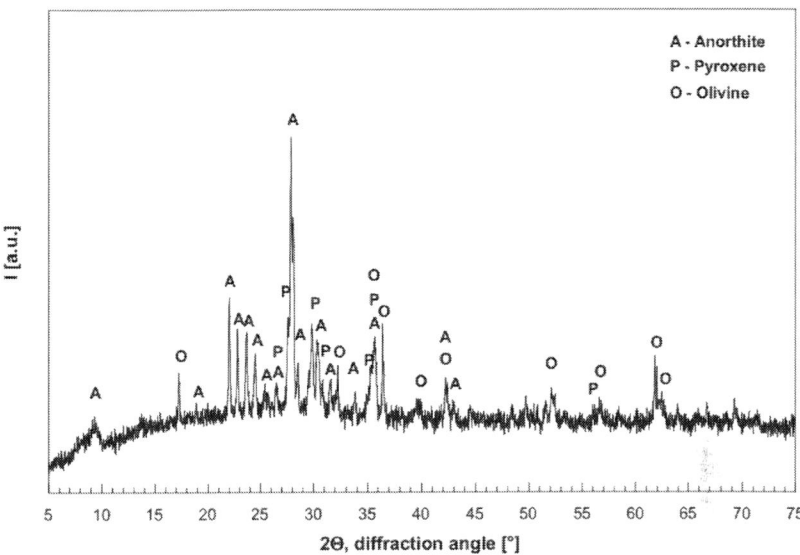

Figure 1: XRD analysis of the as-received Lunar regolith.

Lunar regolith was first sieved, to produce a −45 μm powder fraction, and then blended in suitable proportions with Al (Alfa Aesar, −325 mesh, 99.5% purity) and $FeTiO_3$ (Alfa Aesar, −100 mesh, 99.8% + purity) powders, to obtain the mixtures to be reacted by SHS under either terrestrial or low-gravity conditions. The general reaction stoichiometry is the following:

$$FeTiO_3 + xAl + y(wt.\%)R_L \rightarrow Products \qquad (1)$$

where x was varied in the range 0.9–3, while y was increased from 0 to the maximum allowable weight percentage, depending on the corresponding x value, for obtaining the SHS character in the resulting reacting system. To distinguish mixtures composition, along with the corresponding experimental conditions adopted during SHS, the various systems will be indicated as $S\_x\#\_R_l\#\_P\#\_\#g$, where $x\#$ is the $Al/FeTiO_3$ molar ratio, $R_l\#$ represents the weight percentage (y) of Lunar regolith in the mixture, $P\#$ is the gas pressure level (Torr) inside the reaction chamber, and $\#g$ indicates if the experiment is performed under terrestrial conditions (1 g) or during parabolic flights (0 g).

Each mixture was compacted using an uniaxial press to obtain a series of pellets with cylindrical (11 mm diameter and 25 mm height) or parallelepiped (14 mm × 17 mm × 33 mm) shape.

The apparatus shown in Fig. 2a and b has been specifically designed to perform SHS reactions on the ground and onboard of the Airbus 300 during the 53rd ESA Parabolic Flights Campaign held in Bordeaux (France) last October 2010. The equipment (cf. Fig. 2a) consists of two reaction chambers, each of them containing six holders, a power supply (Legrand safety isolating transformer, mod. 642310, primary 230–400 V, secondary 12–24 V, power 1000 VA) required for reaction ignition and a vacuum pump.

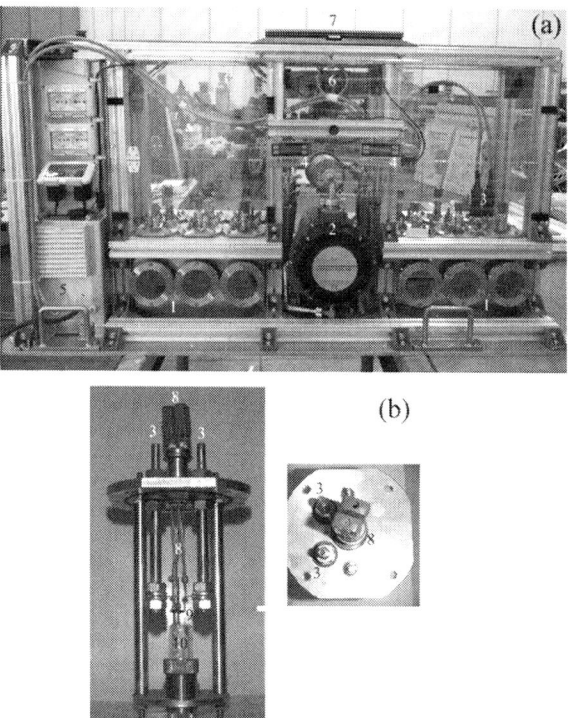

Figure 2: Complete (a) and individual holder (b) views of the apparatus used for SHS experiments performed on the ground and during parabolic flights: (1) reaction chambers, (2) vacuum pump, (3) electrodes, (4) vacuum sensors, (5) power supply, (6) data acquisition system, (7) notebook, (8) thermocouples and related feedthroughs, (9) heating filament, (10) quartz container.

Each individual holder (cf. Fig. 2b) is equipped with a tungsten coil (R.D. Mathis Company, USA) that was electrically heated when the corresponding electrodes were connected to the power supply for activating the SHS reaction. Temperature during reaction evolution was measured using two thermocouples (W–Re, 0.13 mm diameter, Omega Engineering Inc.) equipped with appropriate feedthroughs placed on the top of the holder (cf. Fig. 2b). During terrestrial experiments, a two-color pyrometer Ircon Mirage OR 15-990 (Ircon, USA) was also used for measuring temperature–time profiles of sample surface during SHS evolution. The process was monitored through a video camera (Imaging Source CMOS color camera, model DFK 21AUC03, using Pentax lenses, model B1218A) and a computer (notebook) system connected to a data acquisition board (cDAQ-9174 equipped with NI 9481 high voltage relay module, NI 9213-16ch-24bit thermocouple module and NI 9239-4ch-24bit analog input module, National Instruments) and supported by a software package (LabVIEW, National Instruments).

It should be noted that the video camera above was also employed for the evaluation of the average front velocity, by dividing the sample length by the time required for the reaction to propagate throughout it.

The desired pressure level, in the range 3–760 Torr, was monitored through appropriate vacuum sensors (cf. Fig. 2a).

For systems displaying an SHS behavior, the combustion front generated at the top of the pellet propagates spontaneously through the mixture until it reaches the opposite end of the sample.

The chemical composition of the obtained reacted specimens was determined by X-ray diffraction (XRD) analysis (Philips PW 1830 diffractometer using Cu K Ni-filtered radiation). End products microstructure was examined by Scanning Electron Microscopy (SEM) using an Hitachi S4000 microscope. Compressive strength tests were carried out on SHSed cylindrical specimens using a METRO COM (mod. 100 MI) press equipped with a 10 kN load cell (METIOR CVS).

Parabolic Flight Experiments

The reduced gravity environment was obtained inside a special airbus (Airbus A300) through series of parabolic manoeuvres. Specifically, the aircraft started from a steady normal horizontal flight, and then it took a 1.8 g load factor nosing up to 45° and climbing to 7500 m. Then, the engine thrust was reduced to the minimum required to compensate air-drag. At this point, the aircraft followed a free-fall ballistic trajectory during which weightlessness was achieved. This phase lasted for about 25 s and the peak of the parabola was reached at about 9000 m. At the end of the low gravity period, i.e. again at 7500 m, a symmetrical pull-out phase was executed to bring the aircraft to its steady horizontal flight in about 20 s.

A normal mission lasted about 2 h and consisted of 30 parabolas. During each parabola, gravity was lower than 2×10^{-2} g. Typically, ignition of the SHS reaction was performed at the injection phase (the entry of 0 g phase).

The entire SHS apparatus shown in Fig. 2 was mounted on an aluminum plate provided of suitable holes to permit its fixation to the airbus rails. As indicated in Fig. 2b, a quartz container was used to sustain, during low gravity experiments, the pellet in a stable position.

RESULTS AND DISCUSSION

Selection of the Reducing Agent

A thermodynamic calculation of the adiabatic temperature was conducted recently for mixtures where aluminum or magnesium are considered as possible reducing agents to be directly reacted with JSC Lunar regolith [11]. Based on the relatively higher adiabatic temperature predicted for Mg-based mixtures, it was stated that the latter ones are more convenient, in comparison with the analogous Al-based systems, for being processed by SHS.

However, the low pressure level characterizing Lunar atmosphere plays an important role in the identification of the most suitable reducing metal to be utilized. Indeed, if highly volatile reactants are used, the temperature conditions achieved during the SHS process can lead, unless over-pressured environments are guaranteed, to excessive gas expulsion. Thus, under such circumstances, significant reactants/products losses, sample weight decrease, increasing in product porosity, possibility of the interruption of combustion front propagation, difficulties in process control, etc., likely take place.

In this regard, the effect of pressure on the boiling temperatures of Mg and Al is shown in Fig. 3. On the basis of the reported data, it is clearly seen that Mg is much more volatile than Al so that, the negative phenomena described above may play a relevant role during SHS. This prediction was experimentally confirmed in our study when operating at relatively low pressures (25 Torr) using Mg as reducing agent. Correspondingly, a significant amount of gases was liberated during the SHS process and the resulting product changed significantly its shape and increased its porosity. Moreover, when the pressure level was further decreased, these features led to pellet disintegration.

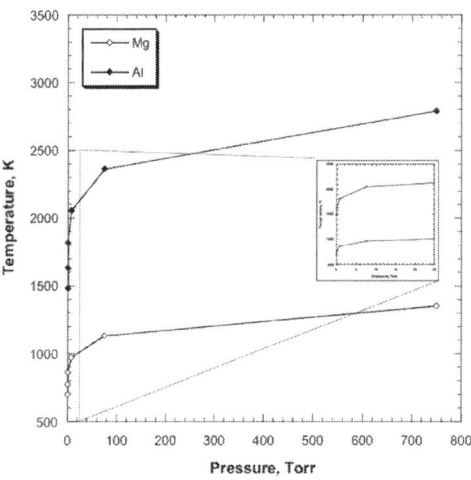

Figure 3: Dependence of boiling temperature on pressure for metal aluminum and magnesium [20].

Consistently with the relatively lower vapor pressure of Al, these undesired phenomena were significantly mitigated when reacting Al-based mixtures. Nevertheless, as shown in Section 3.4, where the experimental results related to the influence of pressure will be reported and discussed, care should be taken also in this case beyond a certain evacuation level.

Based on the considerations above, Al was chosen as reducing metal in our investigation. In this regard, it should be noted that such element could be extracted from Lunar minerals containing it or otherwise recovered from some components or portions of vehicles previously utilized for Space Missions and available on the Moon [11].

Effect of the Al/Fetio$_3$ Molar Ratio

When no regolith was added to the mixtures, all the systems based on reaction (1) exhibited a self-propagating behavior only if $x \geq 0.9$. Moreover, as shown in Fig. 4, when the Al/FeTiO$_3$ molar ratio is augmented, both the average velocity of the combustion front (v_f) and the maximum combustion temperature (T_c) correspondingly increased. This is an indication, at least in the compositional range investigated in this work, of the progressively enhanced exothermic character of reaction (1) with increasing amounts of Al in the system.

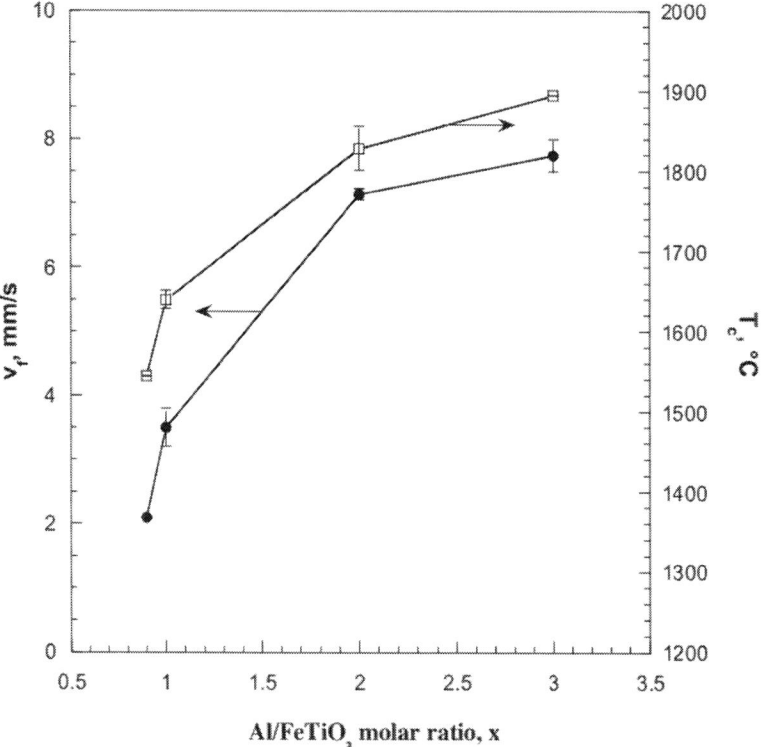

Figure 4: Average front velocity and combustion temperatures as a function of the (Al/FeTiO$_3$) molar ratio in the starting mixture ($y = 0$ in reaction (1)).

The compositions of the resulting products are summarized in Fig. 5.

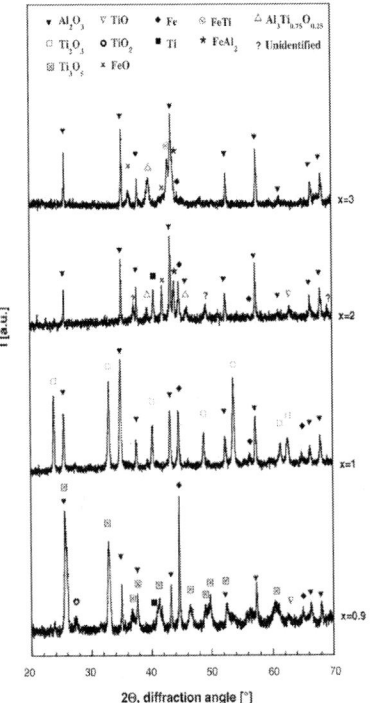

Figure 5: XRD patterns of SHSed products obtained using different *x* values (*y* = 0 in reaction (1)).

Although product composition depends on the system under consideration, all the reported XRD spectra indicate that the SHS process leads to ceramic–metal composite materials with no traces of initial reactants.

Each system will be analyzed separately in what follows, with the aim of identifying the possible global reactions describing the corresponding chemical transformations that take place during the SHS process.

For instance, the formation of the main phases, i.e. Al_2O_3, Fe, and Ti_3O_5, detected by XRD in the final product obtained when using the lowest Al/FeTiO$_3$ ratio (*x* = 0.9), is consistent with the occurrence of the following reaction:

$$FeTiO_3 + 0.9Al \rightarrow 0.45Al_2O_3 + Fe + 0.33Ti_3O_5.$$ ()

The exothermicity of reaction (2), i.e. $(-\Delta H_r^o) = 330.044$ kJ/mol of $FeTiO_3$[21], justifies the self-propagating behavior exhibited by this system.

It should be noted that the secondary species TiO_2 and Ti, also found as minor phases in the SHSed product, are not accounted for in the previous reaction.

When the x value was increased to 1, a relatively less oxidized Ti oxide, namely Ti_2O_3, was formed instead of the Ti_3O_5 phase observed at $x = 0.9$. Moreover, no secondary species were detected in the end-product. These outcomes provide an indication of the higher reducing environment produced by the additional Al. Accordingly, the chemical transformations possibly taking place in this case can be described through the following global reaction:

$$FeTiO_3 + Al \rightarrow 0.5Al_2O_3 + Fe + 0.5Ti_2O_3. \tag{3}$$

The enhanced self-propagating character displayed by the latter one, as compared to system (2) (cf. Fig. 4), is consistent with the corresponding higher enthalpy of reaction (3), i.e. $(-\Delta H_r^o) = 362.753$ kJ/mol of $FeTiO_3$[21].

Analogously, according to the XRD pattern of the SHSed product shown in Fig. 5, the base reaction involved when the $Al/FeTiO_3$ molar ratio is raised to 2 is the following:

$$FeTiO_3 + 2Al \rightarrow Al_2O_3 + Ti + Fe \tag{4}$$

The increase of front velocity and combustion temperature, as compared to the systems with lower xvalues (cf. Fig. 4), is explained also in this case by the relatively higher exothermicity $(-\Delta H_r^o = 440.157$ kJ/mol of $FeTiO_3)$ [21]. It should be noted that, as shown in Fig. 5, the additional presence in the end product of some partially reduced species, particularly FeO, can be associated to the fact that Al not only acts as reducing agent but also reacts with elemental iron to form $FeAl_2$, also detected by XRD.

Thus, the main chemical transformations taking place during SHS for the system with $x = 2$ can be summarized as follows:

$$FeTiO_3 + 2Al \rightarrow (1 - x)Al_2O_3 + Ti + (1 - 4x)Fe + 3xFeO$$
$$+ xFeAl_2 \qquad (5)$$

Nevertheless, other minor species, including some unidentified phases, are not accounted for in this reaction.

The principal new outcome observed as the $Al/FeTiO_3$ ratio was augmented to 3 is the appearance of FeTi at the expenses of elemental Fe and Ti, that tend to disappear from the XRD pattern. In addition, Al_2O_3 remains the main phase detected by XRD while $Al_3Ti_{0.75}O_{0.25}$ and $FeAl_2$ contents increase. Regarding the presence of FeO in the final product, the considerations made previously when examining the case of $x = 2$ are still valid.

The SEM microstructure of a reaction product synthesized under terrestrial conditions when setting $x = 2$ and $y = 0$ in reaction (1) is shown in Fig. 6. According to the corresponding XRD pattern (cf. Fig. 5), the obtained composite material consists of Al_2O_3 grains (dark phase) surrounded by a multiphase metal matrix of Ti, Fe and intermetallic alloys.

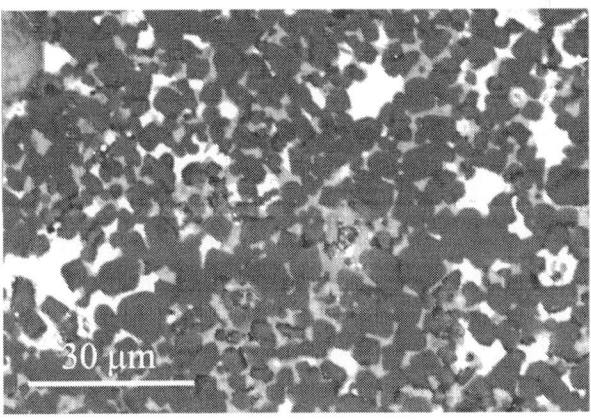

Figure 6: SEM microstructure of the SHSed end-product obtained when using $(Al/FeTiO_3)$ molar ratio equal to 2, no Lunar regolith, gas pressure of 25 Torr, terrestrial conditions ($S\_x2\_R_L0\_P25\_1g$ system).

Effect of the Regolith Content

The addition of Lunar regolith to ilmenite and Al reactants ($y > 0$ in reaction (1)) strongly affects SHS process dynamics of the corresponding reacting systems. In this regard, the influence of regolith content on the measured average front velocity and combustion temperature is plotted in Fig. 7 for the different ($Al/FeTiO_3$) molar ratios investigated. It is clearly observed that both these parameters decrease as the y value is augmented. Moreover, the threshold amount (y_c) of regolith that can be possibly added to the mixture to obtain a self-sustaining reaction is identified. Specifically, above y_c, whose value depends on that of x, the corresponding systems cease to react by SHS. It is also possible to observe from Fig. 7 that the minimum values of front velocity and combustion temperature are in the ranges 1–2 mm/s and 1500–1550 °C, respectively. In particular, the latter condition is perfectly consistent with the range 1800–2000 K that is the empirical requirement for the adiabatic temperature generally reported in the SHS literature for self-sustaining systems [12] and [13].

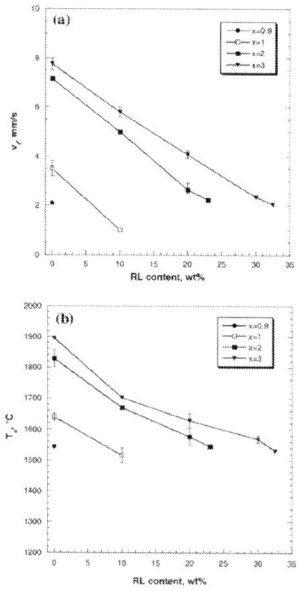

Figure 7: Average front velocity (a) and combustion temperature (b) for the different ($Al/FeTiO_3$) molar ratios investigated as a function of the Lunar regolith content in the starting mixtures.

The composition of the mixtures able to react by SHS using the maximum amount of simulant regolith, that represents the optimal condition in the framework of the ISRU concept, is reported in Fig. 8 as a function of the (Al/FeTiO$_3$) molar ratio. It is clearly seen that, if the regolith content in the mixture is increased, the weight percentage of Al has to be correspondingly augmented for obtaining an SHS reaction.

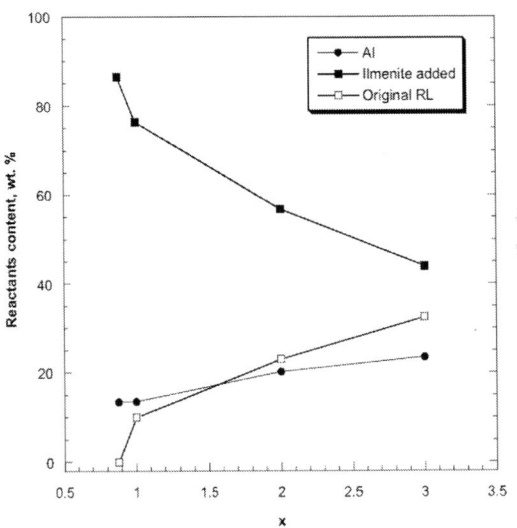

Figure 8: Composition of the optimal reacting mixtures as a function the (Al/FeTiO$_3$) molar ratio used in reaction (1).

Based on these results and considering the percentage of ilmenite present on Moon soil [15], it is possible to estimate the total amount of FeTiO$_3$, as the sum of the fraction originally present in the Lunar regolith and that additionally provided from the external. Moreover, according to the approach followed in a recent paper [10], the sum of these two contributions "simulates" the amount of FeTiO$_3$ in a "modified" Lunar soil, obtained after an enrichment treatment for suitably increasing its ilmenite content. Thus, the resulting enriched soil could be directly reacted with Al.

Consequently, as reported in Fig. 9, it is possible to determine the dependence of the minimum percentage of Al required for making the reactive process self-sustaining as a function of the ilmenite content in the enriched regolith. Regarding this aspect, it should be noted that, although the $FeTiO_3$ amount in certain Lunar basaltic sites could be even higher than 20 wt.% [15], our calculations are based on the assumption that the original content of iron titanate in the regolith is in the range 7–15 wt.%, as obtained from a recent NASA report relatively to Lunar Mare rock types [22].

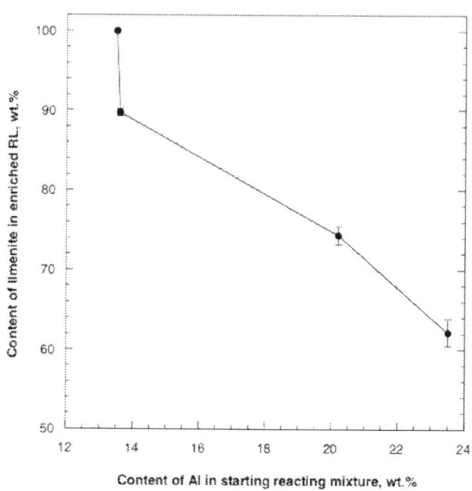

Figure 9: Dependence of the minimum amount of Al required to make the reactive process self-sustaining as a function of the total ilmenite ($FeTiO_3$) content in the enriched regolith.

Fig. 9 indicates that the minimum reducing agent content needed for self-sustaining the process is about 13 wt.% and corresponds to $x = 0.9$ and $y = 0$ in reaction (1). On the other hand, when the maximum amount of Al investigated in this work is added to the mixture, i.e. about 23.5 wt.% (corresponding to $x = 3$), the enrichment stage has to produce a regolith with approximately 61 wt.% of $FeTiO_3$. Thus, depending upon the relative lack of Al available on Moon soil or energy to be spent for the enrichment treatment, different regolith/Al mixtures can be identified for the fabrication of Lunar construction material by SHS.

An example of end-product in the parallelepiped configuration, which could be considered as a possible "Lunar brick" prototype, is shown in Fig. 10 for the case of $S\_x3\_R_L32.5\_P25\_1g$ system. An important issue in view of the obtainment of structural component with selected dimensional characteristics is represented by the fact that the original specimen shape is maintained during the course of the reaction process. This holds also true when considering cylindrical pellets.

Figure 10: Image of the final sample obtained by self-propagating reactive process for the case of parallelepiped configuration, when using ($Al/FeTiO_3$) molar ratio equal to 3, 32.5 wt.% of Lunar regolith in the mixture, gas pressure of 25 Torr, terrestrial conditions ($S\_x3\_R_L32.5\_P25\_1g$ system).

Fig. 11 shows the composition of the products obtained using the different optimal (x, y) combinations identified in the present work (cf. Fig. 8).

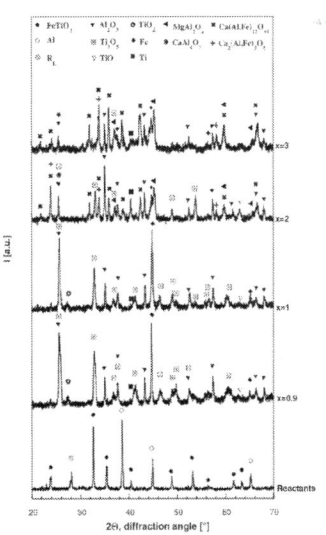

Figure 11: XRD patterns of reactants ($x = 3$, $R_L = 32.5$ wt.%) and SHSed products obtained under the optimal conditions when using different x values.

Analogously to the systems reacted when no regolith was present in the mixture, the starting reagents are also in this case completely converted to multiphase products. Nevertheless, as the Al/FeTiO$_3$ molar ratio is augmented and the maximum amount of regolith allowable for obtaining the SHS behavior in the resulting system is correspondingly increased, a variety of additional phases have been also detected by XRD (cf. Fig. 11). Specifically, as compared to the base case ($x = 0.9$ and $y = 0$), product composition changes only slightly when adding 10 wt.% of regolith to the starting mixture ($x = 1$). Indeed, Al$_2$O$_3$, Fe, Ti$_3$O$_5$, Ti and TiO$_2$ still remain the only phases identified in the end product by XRD. On the other hand, when x 2 and the amount of simulant that is possible to process is correspondingly augmented, the situation varies significantly. Several mixed oxides, namely magnesium spinel (MgAl$_2$O$_4$) and various calcium aluminates (Ca(Al,Fe)$_{12}$O$_{19}$, Ca$_2$(Al,Fe)$_2$O$_5$ and CaAl$_4$O$_7$) are produced during SHS under these circumstances. Other minor or non-crystalline phases, particularly silicates, are also likely formed. The presence of all these species is the result of the chemical interaction of Al with the main constituents initially present in Lunar simulant regolith, i.e. FeTiO$_3$, CaAl$_2$Si$_2$O$_8$, CaSiO$_3$, and Mg$_2$SiO$_4$.

It should be noted that the composition above is also qualitatively similar to that recently reported by Faierson et al. [8] relatively to end-products formed when Lunar regolith simulant was directly reacted with Al (33 wt.% in the mixture). In this case, the reaction was conducted under the so-called volume combustion regime, i.e. after a slowly heating stage (7–15 min) until the mixture achieved the ignition temperature.

An example of SEM micrograph of an SHSed product obtained from mixtures containing Lunar regolith, namely the $S\_x3\_R_l30\_P25\_1g$ system, is shown in Fig. 12. As compared to the case described in Fig. 6, the several phases originally present in the regolith lead to the formation of ceramic–metal composites characterized by relatively complex microstructure. Specifically, the presence of peculiar filament-like structures is well evidenced, particularly inside sample pores (cf. Fig. 12b). Their formation can be likely associated to the occurrence of a vapor–liquid reaction mechanism as a consequence of the Al vaporization taking place during SHS evolution. Unfortunately, EDS analysis did not provide any reliable information regarding the composition of this phase, as we could find them isolated only inside the pores. Nevertheless, since Al is most likely involved in the formation of such structures, they presumably consist of Al-based oxides.

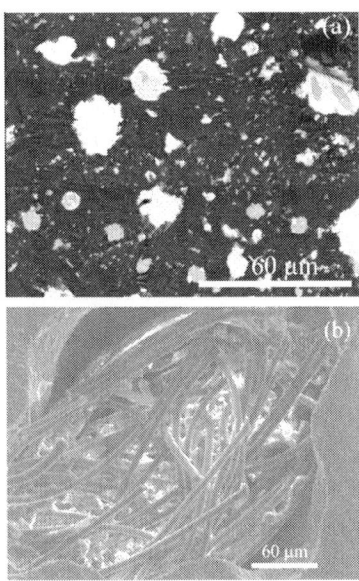

Figure 12: SEM microstructures of the SHSed product obtained when using (Al/FeTiO$_3$) molar ratio equal to 3, 30 wt.% Lunar regolith in the mixture, gas pressure of 25 Torr, terrestrial conditions (S_x3_R$_l$30_P25_1g system): (a) general view of the bulk region; (b) peculiar filament-like structures inside a sample pore.

The presence of whiskers having diameters as small as about 25 nm and probably composed by Al nitrides and oxides was observed by Faierson et al. [8] and Faierson and Logan [9] when performing, under standard (air) conditions, the direct geothermite reaction of Lunar regolith with Al. However, no whiskers were found in this study when the reaction was conducted in a vacuum environment (about 0.6 Torr).

The characterization from the compressive strength point of view of end-products synthesized in the present investigation was conducted on cylindrical S_x2_R$_l$20_P25_1g and S_x3_R$_l$30_P25_1g samples and provided average values of 27.2 ± 3.6 and 25.8 ± 3.6 MPa, respectively. These values are significantly higher in comparison to the best results reported in the literature, i.e. about 18 MPa, relatively to products obtained from the direct aluminothermic reaction of Lunar regolith simulant [8] and [9]. The observed improvement could be ascribed to the increase in reaction exothermicity due to the relatively higher content of ilmenite in the mixture used in the present work. Indeed, such reaction conditions are beneficial for enhancing sintering phenomena during SHS and, consequently, increasing the strength of the resulting material.

Effect of the Pressure Level

All the results reported and discussed in previous sections referred to SHS experiments performed when the evacuation level inside the reaction chamber was 25 Torr. However, since the atmosphere pressure on the Moon is extremely low, i.e. about 10^{-12} Torr (cf. Table 1), while, as discussed in Section 3.1, the SHS process behavior and related product characteristics can be negatively affected when operating under such condition, the influence of this parameter is systematically investigated in this work.

The average front velocity and combustion temperature as a function of the gas pressure inside the SHS chamber are reported in Fig. 13a and b, for the cases of $x = 0.9$ and 2, respectively. Only slight differences are found when decreasing the pressure from 1 atm (Argon environment) down to 25 or 10 Torr. Nevertheless, the situation completely changed when the pressure was further decreased to 3 Torr. Correspondingly, as observed in Fig. 13a and b (see also the related insets), the combustion temperature increases while front velocity decreases. Analogous results are obtained when different $Al/FeTiO_3$ molar ratios were considered.

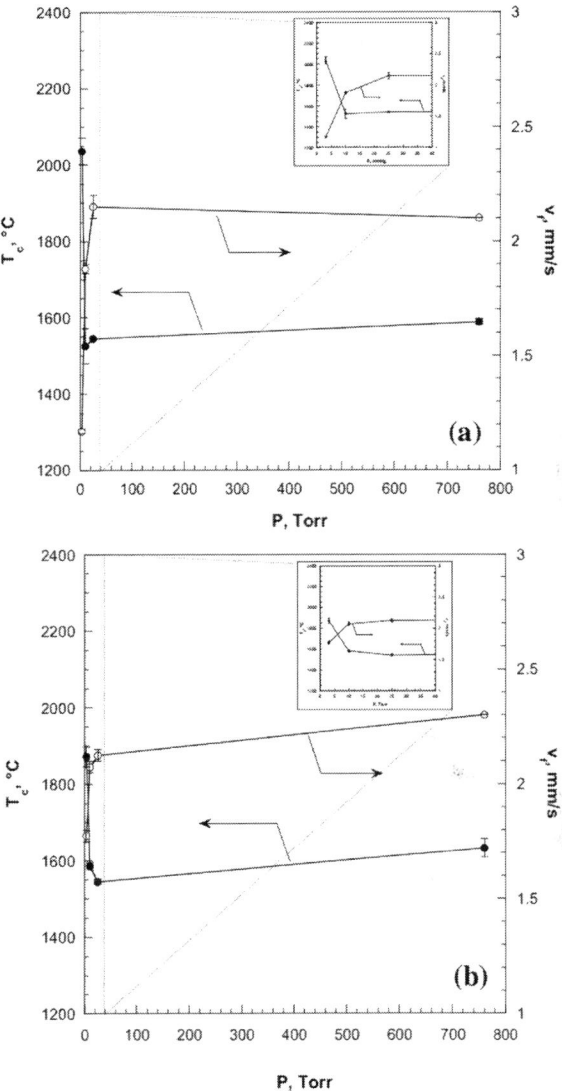

Figure 13: Effect of pressure on combustion temperature and front velocity during self-propagating high-temperature process for the cases of (Al/FeTiO$_3$) molar ratios equal to 0.9 (a) and 2 (b).

These outcomes can be strictly associated to the corresponding weight loss displayed by samples during the SHS process, as evidenced in Fig. 14a and b for the cases of $x = 0.9$ and 2, respectively.

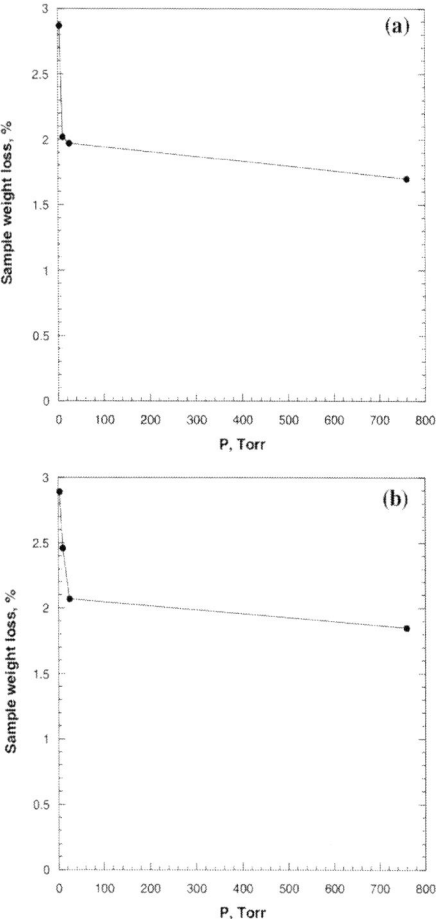

Figure 14: Effect of vacuum level on weight loss in samples obtained by self-propagating high-temperature process for the cases of (Al/FeTiO$_3$) molar ratios equal to 0.9 (a) and 2 (b).

The possible reason for such behavior relates to the fact that, on the basis of the consideration made in Section 3.1, as pressure level is decreased, a relatively larger amount of vapor Al is expected to be developed (cf. Fig. 3). Consequently, part of reactants are progressively expulsed during reaction evolution, thus causing the observed decreasing in sample weight. If the process is conducted at very low pressure, this situation could become critical to lead sample disintegration.

As far as the behavior displayed by wave velocity is concerned, it is likely that the formation of significant amounts of gases may hinder heat transfer phenomena inside the reacting pellets that represent, on the other hand, an important requirement for the progress of the SHS process. Thus, the fact that the propagation speed of the reaction front is lowered finds a possible justification. Nevertheless, the opposite behavior exhibited by the combustion temperature may be associated to the more reactive character displayed by Al when it is in vapor state, that corresponds to the situation encountered when the process is performed at relatively lower pressures.

Based on the consideration above, it is possible to assess that if the fabrication process proposed in this work is carried out under the pressure conditions typically present on Lunar environment, the features discussed previously may play an important role. On the other hand, when the SHS reaction is conducted above a certain pressure threshold (few Torr), no inconvenience will be encountered.

It should be noted that, no remarkable differences in product composition were found within the range of pressure level investigated in the present work.

Effect of Gravity

Due to the low gravity conditions present on the Moon, i.e. about 1.622 m/s^2 (cf. Table 1), another important parameter to be investigated in view of the possible exploitation *in situ* of the process considered in this work is the gravity level under which the self-propagating reaction (1) is conducted.

To evaluate possible differences in process dynamics and/or end-product characteristics as compared to results obtained under terrestrial (1 g) conditions, SHS experiments have been also performed in a microgravity ($\sim 10^{-2}$ g) environment during a recent parabolic flight campaign. Specifically, low-gravity tests have been carried out using Lunar regolith-based systems with $x = 2$ or 3, R_L content in the range 0–30 wt.%, when the pressure inside the chamber was equal to 25 Torr.

The average velocity of the reaction front measured under the two gravitational conditions above are compared in Fig. 15a and b. The sample height was set to 25 mm in order to guarantee that, according to the measured reaction front velocity values (>2 mm/s), the SHS process duration falls entirely within the time interval of about 20 s where the 2×10^{-2} g condition is established during each parabola.

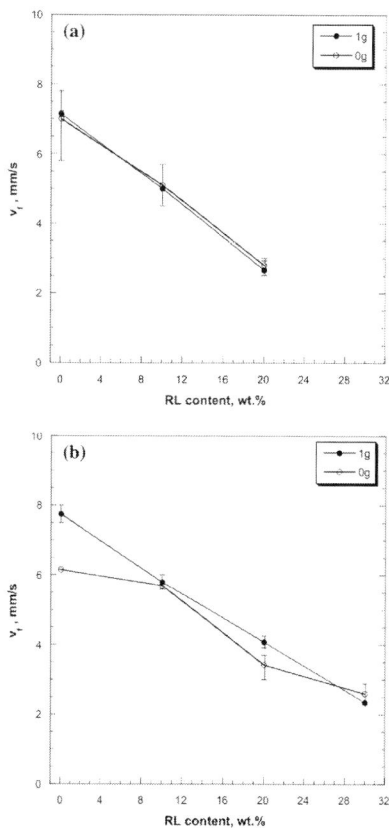

Figure 15: Effect of gravity level on the average velocity of the self-propagating combustion front for the cases of (Al/FeTiO$_3$) molar ratios equal to 2 (a) and 3 (b).

No significant differences are generally displayed when comparing the two situations, except for the case of $x = 3$ and $y = 0$, where the reaction front was observed to propagate relatively slower during experiments conducted under microgravity conditions. A similar finding was reported in the literature when investigating the self-propagating combustion synthesis of TiB$_2$–Ti aluminides based composites at 1 and 2×10^{-2} g levels [23].

The discrepancy observed when $x = 3$ and $y = 0$ can be somehow associated to the larger formation of molten/gaseous phases taking place when processing relatively higher exothermic systems. Indeed, under such circumstances, gravity is expected to mainly affect the process. In any case, this effect tends to vanish when processing the mixtures with compositions similar to the optimal ones identified in the framework of the present work (cf. Fig. 7). Moreover, it should be noted that, since gravity level on the Moon is even higher than that encountered during the 20 s parabolic flight interval, the observed slight differences are expected to be further reduced in Lunar environment.

These considerations hold also true when the comparison is extended to the combustion temperature, although some problems arose with thermocouple measurements, particularly during parabolic flights experiments, whose signals were disturbed when the more exothermic systems were reacted.

Analogous outcomes come also from the compositional point of view and weight loss. In fact, the same phases with similar content were found when characterizing products obtained either under terrestrial or microgravity conditions.

Fig. 16 shows a micrograph of the SHSed sample obtained under low-gravity conditions when setting $x = 2$ and $y = 0$ in reaction (1). Product microstructure is very similar to that resulting when the reaction was performed under terrestrial conditions (cf. Fig. 6). This holds also true when the comparison is extended to systems involving Lunar regolith in the starting mixture.

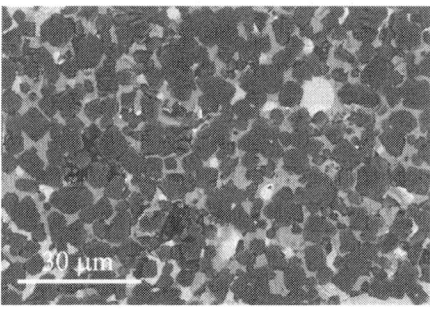

Figure 16: SEM microstructure of the SHSed product obtained when using (Al/FeTiO$_3$) equal to 2, no Lunar regolith, gas pressure of 25 Torr, parabolic flight conditions ($S\_x2\_R_L0\_P25\_0g$ system).

On the basis of the considerations previously made, it is possible to state that gravity plays a marginal role for the SHS systems taken into account in this work. This conclusion is particularly valid when processing relatively low exothermic mixtures, which are, on the other hand, the most interesting from the ISRU point of view.

SUMMARY AND CONCLUDING REMARKS

A recently patented process, based on the exploitation of the self-propagating thermite reduction of ilmenite for ISRU and ISFR applications in Lunar environment, is systematically investigated in this work. Specifically, the dependence of the most important processing parameters, such as the composition of the starting mixture, evacuation level, and gravity conditions, on SHS process dynamics and product characteristics is examined.

Regarding the characteristics of the mixture to be reacted, an important role is played by the choice of the reducing agent, namely magnesium or aluminum. Although mixtures containing magnesium display, as compared to those involving aluminum, a relatively higher self-propagating character [15], the extremely high volatility of the first metal makes the process very difficult to control even at relatively mild evacuation levels. Most important, reacting samples tend to disintegrate during SHS, because of the corresponding gas expulsion. Consequently, the self-propagating aluminothermic reduction of ilmenite was considered as the base reaction of the fabrication process taken into account in this work.

In addition, it is established that the latter process can be performed also using Lunar regolith, that has to be, however, preliminarily enriched in ilmenite, before being reacted with Al. Thus, another relevant aspect relates to the identification of the optimal mixture composition able to guarantee the self-propagating behavior in the resulting system, while limiting the expenses for the enrichment treatment.

In this regard, it is observed that mixtures with $Al/FeTiO_3$ molar ratio lower than 0.9 do not exhibit an SHS behavior while, when this threshold value is set, no additional simulant is allowed to be added to the mixture for obtaining such combustive regime. However, when the amount the reducing metal in the mixture is gradually increased, the SHS process proceeds faster and the measured combustion temperature becomes higher, as a consequence of the increased system exothermicity. Correspondingly, the amount of regolith that is possible to combine with $FeTiO_3$ and Al reactants could be gradually augmented, thus allowing us to identify the optimal mixture composition at the various $Al/FeTiO_3$ molar ratios. This is an important issue in the framework of ISRU applications. Specifically, the minimum amount of Al required by the system for obtaining an SHS reaction increases from about 13 wt.%, for the case of regolith consisting of pure ilmenite, to 23.5 wt.%, needed when the starting mixture contained 32.5 wt.% of simulant. In particular, the latter condition simulates a system consisting of Al with an enriched Lunar soil containing about 62 wt.% of ilmenite. It should be noted that the introduction of the enrichment step leads to a significant decrease in the amount of Al needed by the process, as compared to that (33 wt.%) utilized by Faierson et al. [8].

Regarding end-product composition, Al_2O_3, Fe, and Ti along with, depending upon the $Al/FeTiO_3$ ratio considered, various Ti oxides, are formed in absence of regolith in the initial mixture. The occurrence of possible global reactions is postulated to justify the formation of these phases.

On the other hand, very complex composite ceramic–metal materials, also consisting of additional mixed oxides, such as $MgAl_2O_4$, $Ca(Al,Fe)_{12}O_{19}$, $Ca_2(Al,Fe)_2O_5$ and $CaAl_4O_7$, were produced by SHS when Lunar regolith was present in the starting mixture.

Compressive strength properties of the obtained products are in the range 25.8–27.2 MPa, thus showing a significant improvement as compared to materials produced from the direct thermite reaction of Al with Lunar regolith simulant (10–18 MPa). This outcome is likely associated to the most favorable reaction conditions encountered in our study due to the higher percentage of ilmenite in the initial mixture. Indeed, the corresponding increase in the exothermicity of the reacting system promotes the interaction among reactant particles towards the obtainment of strongly sintered materials.

As far as the effect of vacuum conditions present inside the combustion chamber is concerned, it is found that the SHS process dynamics and product characteristics change only slightly when the pressure level is decreased from 760 down to about 10 Torr. However, the consequences deriving from the enhanced Al vaporization during the process, i.e. sample weight loss, temperature increase and front velocity decrease, become significant when further evacuating the SHS chamber. Thus, in view of the possible exploitation of this fabrication process on the Moon, the synthesis reaction should be conducted in a closed environment at a proper overpressure (few Torr) as compared to the pressure value (10^{-12} Torr) present on Lunar soil.

Finally, the results obtained during parabolic flight experiments, aimed to verify the possible effects caused by the change in gravity when passing from Earth to Moon conditions, reveal that neither SHS process dynamics nor product characteristics are correspondingly influenced in a relevant manner.

All the outcomes reported and discussed in this work allows us to assess that the optimized results obtained under terrestrial conditions are still valid for *in situ* applications in Lunar environment.

ACKNOWLEDGMENTS

The financial support for this work by Italian Space Agency (ASI) under the "COSMIC" Project (Contract number I/033/09/0) is gratefully acknowledged. The authors thank Eng. Monica Valdes (Department of "Ingegneria Strutturale, Infrastrutturale e Geomatica", University of Cagliari, Italy) for performing compressive strength measurements. The European Space Agency (ESA) is also gratefully acknowledged for the 53rd parabolic flight campaign.

REFERENCES

1.   J.A. Bassler, M.P. Bodiford, M.S. Hammond, R. King, C.A. Mclemore, N.R. Hall, M.R. Fiske, J.A. Ray, In Situ fabrication and repair (ISFR) technologies, new challenges for exploration collection of technical papers, in: 44th AIAA Aerospace Sciences Meeting, vol. 6, 2006, pp. 4166–4172.

2.   M.S. Hammond, J.E. Good, S.D. Gilley, R.W. Howard, Developing fabrication technologies to provide on demand manufacturing for exploration of the Moon and Mars collection of technical papers, in: 44th AIAA Aerospace Sciences Meeting, vol. 9, 2006, pp. 6353–6360.

3.   J.T. Howell, J.C. Fikes, C.A. McLemore, J.E. Good, On-site fabrication infrastructure to enable efficient exploration and utilization of space, in: International Astronautical Federation – 59th International Astronautical Congress 2008, vol. 12, 2008, pp. 7842–7848.

4.   C. Allen, J. Graf, D. McKay, Sintering bricks on the moon, engineering, construction, and operations, Space IV Am. Soc. Civ. Eng. (1994) 1220–1229.

5.   H. Toutanji, B. Glenn-Loper, B. Schrayshuen, Strength and durability performance of waterless Lunar concrete, in: 43rd AIAA Aerospace Sciences Meeting and Exhibit – Meeting Papers, 2005, pp. 11427–11438.

6.   D. Tucker, E. Ethridge, H. Toutanji, Production of glass fibers for reinforcing Lunar concrete, in: Collection of Technical Papers, 44th AIAA Aerospace Sciences Meeting, vol. 9, 2006, pp. 6335–6343.

7.   K.S. Martirosyan, D. Luss, Combustion synthesis of ceramic composites from Lunar soil simulant, in: 37th Lunar and Planetary Science Conference (2006) Abstract, 1896.

8.   E.J. Faierson, K.V. Logan, B.K. Stewart, M.P. Hunt, Demonstration of concept for fabrication of Lunar physical assets utilizing Lunar regolith simulant and a geothermite reaction, Acta Astronaut. 67 (2010) 38–45.

9.   E.J. Faierson, K.V. Logan, Geothermite reactions for in situ resource utilization on the moon and beyond, in: Proc. of Earth and Space 2010: Engineering, Science, Construction, and Operations in Challenging Environments, 2010, pp. 1152–1161.

10.   G. Corrias, R. Licheri, R. Orrù, G. Cao, Self-propagating high-temperature reactions for the fabrication of Lunar and Martian physical assets, Acta Astronaut. 70 (2012) 69–76.

11.   C. White, F. Alvarez, E. Shafirovich, Combustible mixtures of Lunar regolith with aluminum and magnesium: thermodynamic analysis and combustion experiments, AIAA (2011–613) (2011) 11.

12.   Z.A. Munir, U. Anselmi-Tamburini, Self-propagating exothermic reactions: the synthesis of high-temperature materials by combustion, Mater. Sci. Rep. 3 (1989) 279–365.

13.   A. Varma, A.S. Rogachev, A.S. Mukasyan, S. Hwang, Combustion synthesis of advanced materials: principles and applications, Adv. Chem. Eng. 24 (1998) 79–226.

14.   G. Cao, A. Concas, G. Corrias, R. Licheri, R. Orrù, M. Pisu, C. Zanotti, Fabrication Process of Physical Assets for Civil and/ or Industrial Structures on the Surface of Moon, Mars and/or Asteroids, Patent 10453PTWO, Applicant, Università di Cagliari and Italian Space Agency, Italy, 2011.

15.   D. Schrunk, B. Sharpe, B.L. Cooper, M. Thangavelu, The Moon Resources, Future Development and Colonization, second ed., Springer-Verlag Inc., New York, 2008.

16.   Y. Wu, C. Yin, Z. Zou, H. Wei, X. Li, Combustion synthesis of fine TiFe series alloy powder by magnetothermic reduction of ilmenite, Rare Met. 25 (2006) 280–283.

17.   Z. Zou, Y. Wu, C. Yin, X. Li, Preparation of Fe–Al intermetallic/ TiC–Al2O3 ceramic composites from ilmenite by SHS, J. Wuhan Univ. Technol. Mater. Sci. Ed. 22 (2007) 706–709.

18.   J.J. Moore, H.C. Yi, J.Y. Guigné, The application of self-propagating high temperature (combustion) synthesis (SHS) for in situ fabrication and repair (ISFR), and in situ resource utilization (ISRU), Int. J. Self-Propag. High Temp. Synth. 14 (2005) 131–149.

19.   G.H. Heiken, D.T. Vaniman, B.M. French, The Lunar Sourcebook – A User's Guide to the Moon, Cambridge University Press, 1993.

20.   D.R. Lide, CRC Handbook of Chemistry and Physics, Eightieth ed., CRC Press, 1999.

21.   I. Barin, Thermochemical Data of Pure Substances, VHC, Weinheim, 1989.

22.   L. Sibille, P. Carpenter, R. Schlagheck, R.A. French, Lunar regolith simulant materials: recommendations for standardization, production, and usage, NASA Technol. Rep. (2005).

23.   A.M. Locci, R. Licheri, R. Orrù, A. Cincotti, G. Cao, J. De Wilde, F. Lemoisson, L. Froyen, I.A. Beloki, A.E. Sytschev, A.S. Rogachev, D.J. Jarvis, Low-gravity combustion synthesis: theoretical analysis of experimental evidences, AIChE J. 52 (11) (2006) 3744–3761.

Effect of the Constituents (Asphalt, Clay Materials, Floating Particles and Fines) of Construction and Demolition Waste on the Properties of Recycled Concretes

C. Medina[a], W. Zhu[b], T. Howind[b], M. Frías[c], and M.I. Sánchez de Rojas[c]

[a]School of Engineering, UEX-CSIC Partnering Unit, University of Extremadura, Avda. De la Universidad, s/n, 10071 Cáceres, Spain

[b]School of Engineering, University of the West of Scotland, Paisley Campus, Paisley PA1 2BE, United Kingdom

[c]"Eduardo Torroja" Institute for Construction Sciences, C/Serrano Galvache, 4, 28033 Madrid, Spain

ABSTRACT

The present study explores the viability of reusing mixed recycled aggregate from construction and demolition waste as a partial (25 and 50 wt%) replacement for natural coarse aggregate in the manufacture of concretes with a compressive strength of 30 MPa. It further analyses the effect of some of the constituents (asphalt, clay-based materials, floating particles and fines) of these recycled aggregates on the properties of recycled concretes. Despite the high asphalt and floating particle content of the recycled aggregate used, came from waste management plant at Glasgow, it was found to have no adverse effect on the workability of the new concretes. Hardened concrete density and compressive strength were observed to decline with increasing replacement ratios, at a variable rate depending on the components of the recycled aggregate mix and the thickness of their ITZs (the thicker the weaker). While concrete with 25% recycled aggregate exhibited lower sorptivity than the reference concrete, absorption was higher when the replacement ratio was 50%. The findings showed that this type of recycled aggregate can be used in concrete manufactured for housing applications and confirmed the importance of good construction and demolition waste management to deliver high quality recycled aggregate.

INTRODUCTION

Concrete is the construction industry's most popular material because of its mechanical properties, durability, cost effectiveness and availability. Concrete output in the European Union (EU-27) is estimated to be on the order of $1350 \cdot 10^9$ t/year [1].

Such a sizeable volume entails the consumption of vast amounts of natural resources (aggregate) that could be replaced by recycled materials such as aggregates from construction and demolition waste (C&DW). Approximately $3 \cdot 10^{12}$ tonnes of (coarse and fine) aggregate are produced yearly in the EU-27[2].

Scotland's 9 million tonnes of C&DW account for over 44% of the country's total annual waste. At this time, 75% of Scots C&DW is reused. That value is higher than the European average (43%) but lower than found in countries (>80%) such as Netherlands, Denmark, Estonia and Germany [1]. Valorisation lowers the percentage of waste deposited in landfills, yields by-products that can be used as prime materials in construction, and furthers sustainable development by reducing CO_2 emissions and the exploitation of natural resources [3].

C&DW is treated at specific plants (Fig. 1) where, after an initial inspection, it is subjected to a series of operations (initial screening, crushing, magnetic separation, manual separation of impurities, mechanical grinder, etc.) that may vary depending on the initial composition of the waste and the end product requirements.

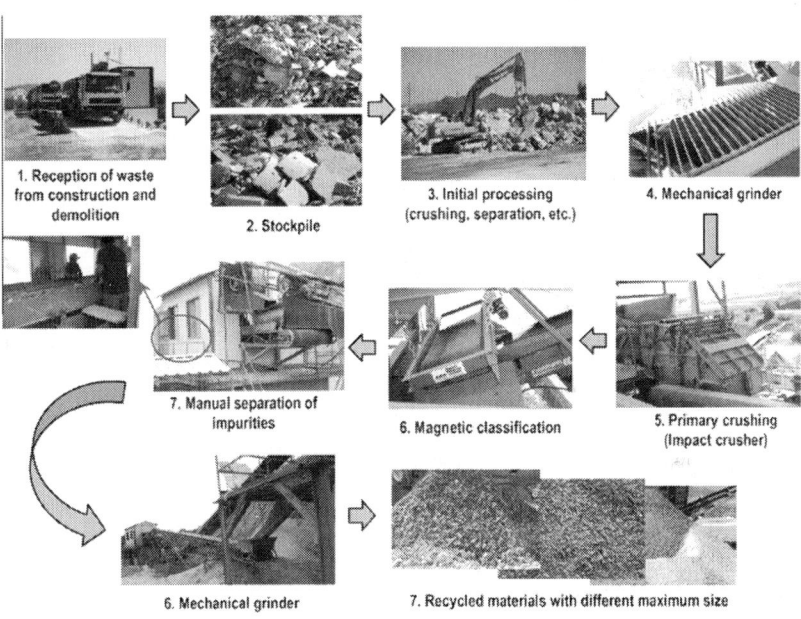

Figure 1: Recycling process of CDW.

Depending on its constituents, recycled aggregate (the end product) is divided in British standard BS 8500-2 [4] into two classes: recycled concrete aggregate (RCA), containing essentially crushed concrete (≥ 95%), and mixed recycled aggregate (RA), comprising stone-based materials (such as concrete, bricks, roof tiles or asphalt), as well as organic (including wood, plastic and cardboard) and inorganic (metal and gypsum plaster) matter.

The volume of RCA generated yearly in Europe is smaller than the amount of RA [5]; in Spain, for instance, an estimated 67% of the total recycled aggregate is RA [6].

Recycled concrete aggregate has been the focus of countless research studies [7], [8], [9] and [10]. As a rule, the use of this material as a partial (<50%) replacement for natural coarse aggregate has been reported to induce minor variations in the physical and mechanical properties of the resulting concretes, due to its intrinsic properties (sorptivity, density, shape and texture).

The application of mixed recycled aggregate (RA) from C&DW to manufacture concrete has been the object of much less research [6], [10], [11], [12] and [13], because of the complexity deriving from the variability of the material involved. Depending on the sources of the waste and the recycling technology, the resulting RA may contain impurities (such as wood, plastic, clay-based materials or asphalt) whose presence affects fresh and hardened concrete performance adversely. Previous research has shown that RA-containing concrete has lower density and poorer physical (workability) and mechanical properties than conventional concrete. The decline in concrete compressive strength usually intensifies with increasing RA replacement ratios, with slides as steep as 30% in concrete made with 100% of RA.

A few researchers have nonetheless studied the effect of recycled aggregate components (such as clay-based materials, asphalt and floating particles) on the end product [14], [15] and [16]. These authors analysed the impact of the presence of aluminium, plastic and different percentages of clay-based materials on the physical and mechanical properties and durability of the new concretes, respectively.

The present study explored the effect of the main constituents (asphalt, clay-based materials, floating particles and fines) of recycled aggregate on the properties of 30 MPa recycled concretes. The RA used in this study was provided by a construction and demolition waste management plant in Glasgow, Scotland. The study aimed to use that aggregate (with or without floating particles, asphalt, brick and fines) to replace 25% or 50% of the natural coarse aggregate in concrete apt for housing applications. The research focused on the effects of the RA replacement ratios and the removal of impurities on the resulting concrete density, workability, sorptivity and compressive strength and the microstructure of the coarse aggregates/paste ITZs.

MATERIALS AND EXPERIMENTS

Materials

The natural coarse and fine aggregate (Fig. 2) used was crushed siliceous rock. The morphology of the coarse fraction (gravel) was irregular and rough, with sharp edges and a maximum size of 20 mm, while the fines (sand) comprised particles of under 4 mm. Its chemical composition (Table 1) consisted primarily of silica and aluminium oxides (67 wt%), with smaller proportions of other oxides (Fe_2O_3, Na_2O, CaO) and a number of trace elements. While quartz prevailed in its mineralogy, feldspars and phyllosilicates were also present.

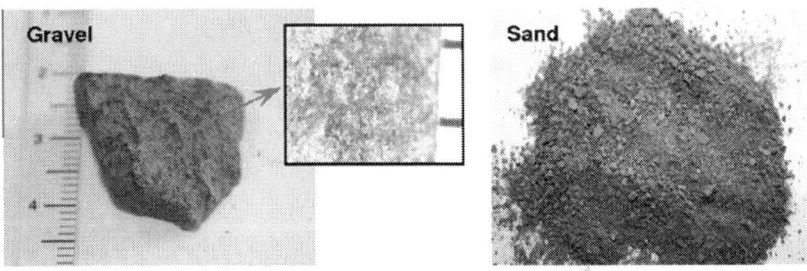

Figure 2: Natural aggregate.

Table 1: Chemical composition of sand, gravel, RA and constituents

Element/ compound (weight %)	Sand	Gravel	RA	Clay-based materials	Concrete waste	Unbound aggregate
SiO_2	66.49	51.26	54.37	76.27	58.29	69.30
Al_2O_3	14.76	16.36	12.90	11.91	7.69	9.32
Fe_2O_3	5.13	6.47	8.28	3.64	6.12	6.99
CaO	3.64	2.96	7.94	2.31	13.27	3.62
Na_2O	2.42	8.25	2.78	0.47	1.45	2.52
K_2O	1.88	4.57	1.68	2.33	0.80	1.37
MgO	0.81	2.39	4.07	0.97	2.28	2.15
TiO_2	0.36	1.30	1.44	0.00	0.00	0.11

SO$_3$	0.14	0.00	0.24	0.66	0.92	1.23
P$_2$O$_5$	0.13	0.44	0.28	0.00	0.28	0.03
MnO	0.07	0.19	0.15	0.09	0.16	0.24
BaO*	0.39	0.09	0.09	489.25	670.52	723.34
Sr*	324.62	489.84	345.85	101.52	414.98	293.15
Zr*	154.49	301.76	166.96	699.17	–	–
Cl*	–	441.75	563.82	–	179.49	–
Rb*	–	191.41	179.57	–	–	–
Zn*	–	172.61	130.52	327.14	–	102.87
Ni*	–	55.23	142.68	–	107.13	85.14
Cr*	–	130.51		–	206.75	196.77
LOI	3.71	5.39	5.52	1.15	8.51	2.93

*In ppm.

The recycled aggregate (RA) used was a typical product from a C&DW management and processing plant in Glasgow, Scotland (Fig. 1). With a maximum particle size of 20 mm, it exhibited compositional heterogeneity and a morphology that varied depending on the source material (see Fig. 3 and Fig. 5). The material selected was analysed to determine its physical, chemical and mechanical composition.

Figure 3: Mixed recycled aggregate.

The Portland 52.5 R cement used was standard BS EN 197-1-compliant [17].

Concrete Design

The following concrete mixes were studied for the present study: a reference concrete (RC); two recycled concretes, one with 25% and the other 50% recycled aggregate, with (RCF-25 and RCF-50) or without (RC-25 and RC-50) floating particles; two recycled concretes, one with 25% and the other 50% recycled aggregate, without (RC25-AS and RC50-AS) asphalt; two recycled concretes, one with 25% and the other 50% recycled aggregate, without (RC25-BR and RC50-BR) clay-based materials; and two recycled concretes with 50% recycled aggregate, with or without floating particles and without fines (RCF-50-ii and RC-50-ii, respectively).

The mixes were proportioned as specified in the *DOE British Method* [18], to a characteristic strength of 30 MPa and w/c ratio of 0.60.

The amount of mixing water added was adjusted to take the water absorption of the aggregates (natural aggregates and RA).

The mix proportions for each concrete mix are given in Table 2.

Table 2: Concrete mix proportions and slump tests results

Concrete	Material (kg/m³)						Cement	Water	(w/c) effective	(w/c) apparent	Slump (cm)
	Sand	Gravel	RA*	RA_WF**	RA_AS***	RA_BR****					
RC	953.72	1033.20	–	–	–	–	323.08	210.00	0.65	0.65	32
RCF-25	948.92	771.00	259.39	–	–	–	323.08	219.75	0.65	0.68	32
RCF-50	941.72	510.10	514.84	–	–	–	323.08	229.36	0.65	0.71	32
RCF-50-ii^a	941.72	510.10	514.84				323.08	229.16	0.65	0.71	35
RC-25	948.92	771.00	–	259.39			323.08	219.54	0.65	0.68	35
RC-50	941.72	510.10	–	514.84			323.08	228.94	0.65	0.71	32
RC-50-ii^b	941.72	510.10	–	514.84			323.08	228.27	0.65	0.71	34
RC25-AS	948.92	771.00	–		259.39		323.08	219.35	0.65	0.68	40
RC50-AS	941.72	510.10	–		514.84		323.08	228.76	0.65	0.71	35
RC25-BR	948.92	771.00	–			259.39	323.08	218.75	0.65	0.68	43
RC50-BR	941.72	510.10	–			514.84	323.08	227.68	0.65	0.71	40

aConcrete prepared with recycled aggregate containing floating particles but no fines (RA_ll).

bConcrete prepared with recycled aggregate containing neither floating particles no fines (RA_WFll).

*Recycled aggregate containing floating particles.

**Recycled aggregate containing no floating particles.

***Recycled aggregate containing no asphalt.

****Recycled aggregate containing no brick.

The grading (Table 3) of the aggregate mix (sand, gravel and mixed recycled aggregate) was practically constant in all the concretes tested to obviate the possibility that alterations in their physical or mechanical properties could be attributed to size distribution differences.

Table 3: Grading of the aggregate mix

Sieve (mm)	RC	Concretes with 25% RA	Concretes with 50% RA
31.5	100.00	100.00	100.00
20.0	99.91	99.60	99.29
16.0	97.42	94.89	92.36
14.0	92.70	89.13	85.56
10.0	79.12	75.41	71.70
8.0	71.05	68.90	66.76
6.3	63.87	63.37	62.88
4.0	53.66	55.54	57.41
2.0	36.35	38.67	40.98
1.0	20.37	22.52	24.66
0.5	11.91	13.73	15.56
0.25	6.41	8.24	10.06
0.125	2.95	4.78	6.61
0.063	1.05	1.19	1.34

The mixes prepared conformed to the minimum cement content and maximum w/c ratio values specified in Table A.14 in British standard BS 8500-1 [19] for concrete to be used in housing and other applications.

Characterisation of the Recycled Mixed Aggregates and Concretes

The mixed recycled aggregate was characterised by determining first its components (such as asphalt and clay-based materials) and subsequently its chemical, physical and mechanical properties (Table 4).

Table 4: Characterisation of the aggregates and concrete mixes

Material	Property	Standard
Aggregates	Constituents of coarse recycled aggregates	BS EN 933-11
	Particle size distribution	BS EN 933-2
	Density and water absorption	BS EN 1097-6
Concrete	Slump test	BS EN 12350-2
	Density	BS EN 12390-7
	Compressive strength	BS EN 12390-3
	Sorptivity	

Aggregate chemical composition was determined on a Bruker S8 TIGER wavelength-dispersive X-ray fluorescent spectrometer, using QUANT EXPRESS standardless calibration (SEPECTRAPlus package) software. Its mineralogy was analysed by powder X-ray diffraction (XRD) on a BRUKER Theta–Theta D8 Advance 2.2-kW Cu anode, non-monochromator spectrometer. Readings were recorded at 2 theta diffraction angles ranging from 5 to 60°, with a step size of 0.019° and a count time per step of 0.5 s.

Table 4 also lists the physical and mechanical properties of the concretes prepared, determined on ninety-nine 100 mm^3 specimens (nine specimens/concrete type): 66 to assess compressive strength and the rest to determine water absorption.

Lastly, the microstructure of the ITZs between the (natural and recycled) coarse aggregate/paste was explored with BSE–EDX. These microstructural studies were conducted by spot chemical analysis using a HITACHI model S-4800 scanning electron microscope; tungsten source energy-dispersive X-rays; and silicon detector. X-ray line scanning analyses (10 lines/ITZ) were recorded in the ITZs between the coarse aggregate constituents (gravel, asphalt, clay-based materials and floating particles) and the paste to determine variations in the Ca/Si ratio. An example of the analysis is given in Fig. 4. The count rate for each analysis obtained was 300 s. The samples were epoxy-coated and precision-sawed and their flat surfaces were carefully polished for backscattering electron (BSE) microscopic analysis, which was conducted to identify their microstructure.

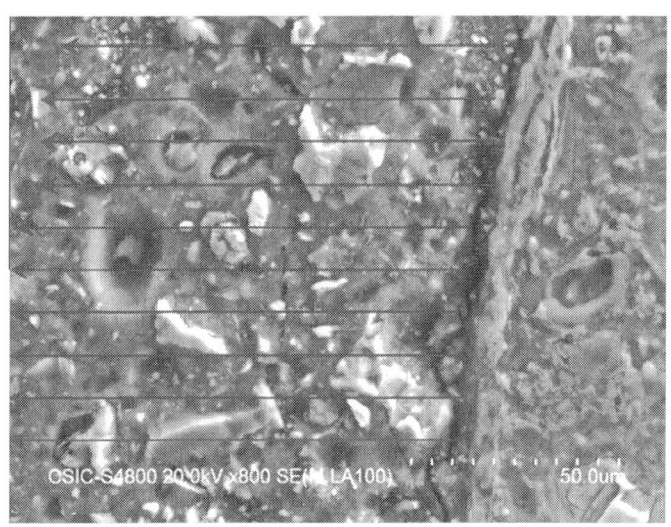

Figure 4: X-ray line scanning analyses in the ITZ.

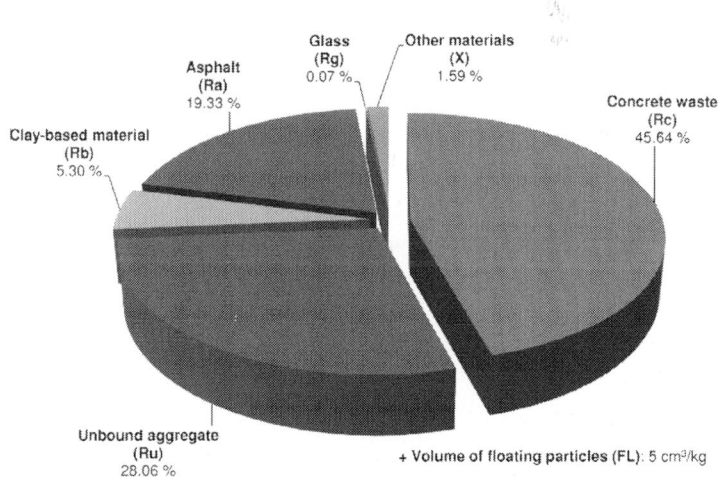

Figure 5: Percentage breakdown of recycled aggregate constituents.

RESULTS

Characterisation of the Ra

Compositional Characterisation

Fig. 5 shows the weight percentage of the recycled aggregate constituents, which included primarily stone-based (Rc > Ru > Ra > Rb) but also minority materials (such as glass, metal and wood).

Further to the classification proposed by Agrela et al. [13], the RA used was a mixed recycled aggregate, for it had a clay-based material content of 0–30% and a concrete and natural aggregate content (Rc + Ru) of 70–90%.

The RA studied conformed to British standard BS 8500-2 [4] and BS EN 12620 [20] provisions on the percentage of clay-based material allowed in aggregates for recycled concrete manufacture. The proportion of asphalt (Ra) and other materials (X + Rg + Floating particles), however, exceeded the 10% and 1% limits, respectively, established in those standards.

Additionally, the percentage of asphalt (Ra) obtained is too higher than the specifications established for German standard DIN 4226-100 ($\leq 1\%$) [21], Laboratório Nacional Portuguese de Engenharia Civil E-471 ($\leq 5\%$) [22] and draft of European Committee for Standardization CEN/ TC 104/SC ($\leq 5\%$) [23] for this type of recycled aggregate. However, the amount of X + Rg exceeded the limit indicated for Portuguese standard E-471 ($\leq 0.5\%$) and is lower than 2% established in draft of CEN/TC 104/SC. Finally, the amount floating particles (FL) is lower than the corresponding value according to the draft of European standard (≤ 2 cm^3/kg).

The morphology of the mixed recycled aggregate components is illustrated in Fig. 6: the stone-based materials were characterised by a rough texture, irregular shape and sharp edges, while the fraction containing the other materials and floating particles exhibited a greater diversity of shapes (round, elongated, flat...) due to the variety of the components.

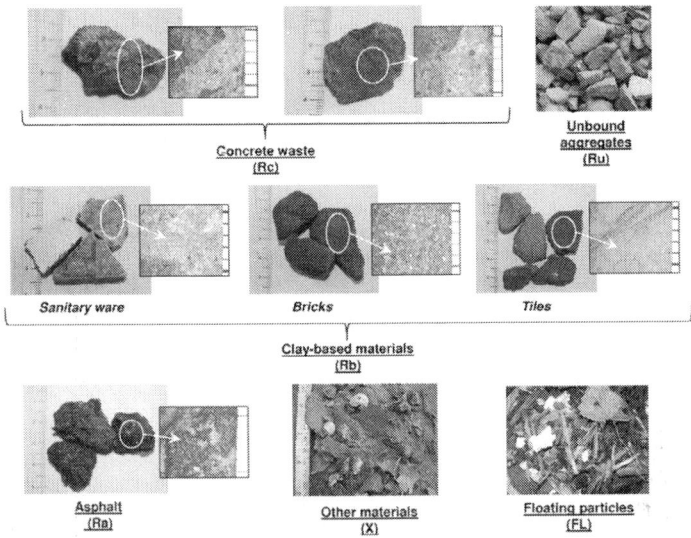

Figure 6: Mixed recycled aggregate constituents: images.

The concrete waste (Rc) consisted of three readily distinguishable parts: coarse aggregate, attached mortar and a coarse aggregate/mortar interface. The latter two induced higher porosity, greater water sorptivity and higher Los Angeles coefficients in the RCA than in natural aggregate [24] and [25]. The clay-based fraction, in turn, comprised two groups with differing morphology and physical and mechanical properties: (a) reddish colour, variably shaped (round or platy) bricks and roof tiles which exhibited higher water absorption and porosity and lower resistance to fragmentation than natural aggregate [26]; and (b) sanitary ware waste, with two clearly distinguishable, differently coloured parts (inner matrix and outer enamel coating), a flat, platy shape, higher porosity and water sorptivity and a lower Los Angeles coefficient than natural aggregate [27]. Finally, the characteristically black asphalt fraction was readily identifiable by that feature and its rough exterior and rounded shape.

Chemical and Mineralogical Composition

The chemical composition of the C&DW and some of its constituents (clay-based materials, concrete waste and unbound aggregates) is listed in Table 1.

The Table 1 shows that the main compounds in all the constituents were silicon oxide (>54 wt%), aluminium oxide (>7.5 wt%), iron oxide (>3.5 wt%) and calcium oxide (>2.3 wt%), along with other minority oxides. Trace elements such as Ba, Zr, Cl and Cr were also found (shown in ppm).

The majority elements in the asphalt fraction, carbon (83 wt%) and hydrogen (8 wt%), were found along with smaller percentages of nitrogen, sulphur and oxygen as well as trace elements (Si, Al, Fe).

The chemical composition of the RA (C&DW) showed values of SiO_2, Al_2O_3, CaO and LOI that lay within the range reported by other authors [28] and [29]. Nonetheless, the Fe_2O_3 values were higher than observed in other similar waste [30], inasmuch as this oxide was present in some of the components (concrete waste and unbound aggregate).

Fig. 7 shows the mineralogical composition of the RA and its stone-based components. Concrete waste, unbound aggregate and clay-based materials were identified. The most abundant mineral in the RA proved to be quartz, although feldspars (albite, orthoclase and sanidine), hematites, magnetite and calcite were also present. These findings were consistent with the mineral phases identified by other author [5],[31] and [32].

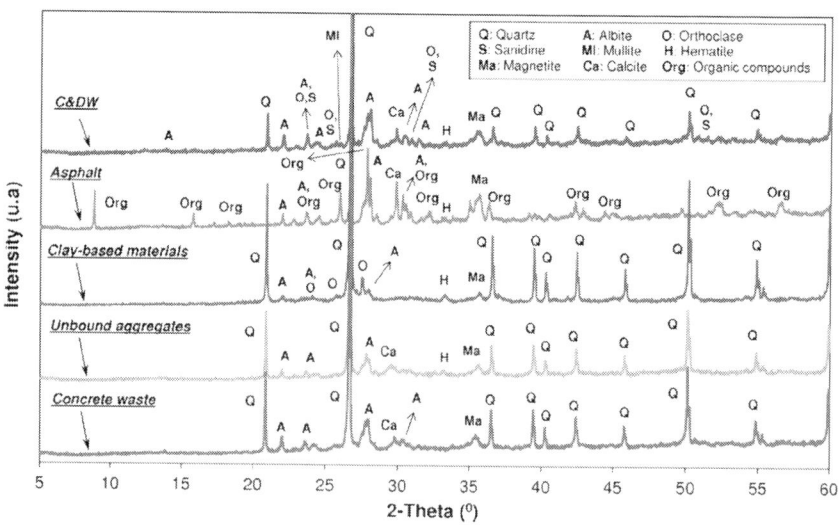

Figure 7: Mineralogical composition of RA (C&DW) and constituents.

The asphalt was observed to contain a majority of organic minerals (such as malonamide and ammonium copper malonate), as well as the minerals present in its aggregate component, such as quartz, albite, magnetite and calcite.

Physical Properties of the Natural and Recycled Aggregates

Results of the physical properties of the RA are presented in Table 5. The grading curves of the aggregates mix in each concrete studied are shown in Table 3. The maximum particle size of both the recycled and the natural aggregate was 20 mm (see Table 5), as may be deduced from the particle size distribution showed in Table 2.

Table 5: Physical properties of the aggregates

Property	Aggregate type					
	Sand	Gravel	RA*	RA$_{WF}$**	RA$_{AS}$***	RA$_{BR}$****
Maximum size (mm)	4	20	20	20	20	20
Particle density on a saturated and surface-dried base (kg/dm³)	2.62	2.66	2.54	2.56	2.56	2.58
Water absorption (% wt.) for 24 h	1.07	2.66	4.49	4.36	4.45	4.42

*Recycled aggregate containing floating particles.

**Recycled aggregate containing no floating particles.

***Recycled aggregate containing no asphalt.

***Recycled aggregate containing no brick.

The results showed that the recycled aggregates (RA) were less dense than the natural material, by 4.49%, 1.99%, 3.87%, and 3.12% for RA, RA$_{WF}$, RA$_{BR}$ and RA$_{AS}$, respectively. Such lower density was attributed to the (less dense) mortar attached to the aggregate and the lower density of the asphalt and clay-based and floating materials and concurred with earlier reports [33] and [15]. The value obtained lay within the 2.22–2.58 Mg/m^3 range found in previous studies on mixed recycled aggregate [10] and [13].

The mixed recycled aggregates (RA, RA_{WF}, RA_{BR} y RA_{AS}) were observed to absorb 1.68, 1.63, 1.60 and 1.69 times more water than the gravel, due mainly to the higher porosity of the attached mortar [8] and the clay-based materials [27], [34] and [35]. The absorption values for this mixed recycled aggregate (RA) are within the range (2.1–8.79%) reported for other types of recycled mixed aggregates [13].

Table 4 reveals the effect of the various stone-based components on water absorption, which was more intense in RA_{AS} than in the other recycled aggregates (RA, RA_{WF} and RA_{BR}) because the component removed (asphalt) had a low water absorption coefficient [36]. Removing the floating particles (RA_{WF}) and clay-based material (RA_{BR}) led to greater declines in the water absorption coefficient because water sorptivity is high in both clay materials and floating particles.

Finally, the absorption values for these mixed recycled aggregates were below the 5% ceiling laid down in the Spanish Code on Structural Concrete [37].

Properties of Concretes

Slump of Fresh Concretes

The slump test findings for the concretes are shown in Table 2.

As observed in earlier studies [16], the inclusion of mixed recycled aggregate had no adverse effect on the workability of the new concretes, irrespective of the replacement ratio or type of mixed recycled aggregate. The aggregate moisture and water absorption values were factored into the calculations for the amount of mixing water needed to sidestep the problems in this respect reported by other authors [11] and [12]working with these materials.

The figure likewise shows that recycled concretes RCF-50-ii and RC50-ii were more workable than concretes RCF-50 and RC-50 because the component removed from the RA (fines), had a large specific surface and high water absorption. Removing the asphalt and clay-based fractions was also observed to induce greater workability, although less notably at higher replacement ratios.

Density of Hardened Concretes

Table 6 gives the density values for saturated concretes at 7 and 28 days, which were observed to rise with age as a result of cement hydration and the concomitant decline in porosity. These density values were within the range (2.4–2.13 kg/dm³) observed previously in concrete containing aggregate sourced from C&DW [6].

Table 6: Density of the concrete mixes

Concrete mix	Density (kg/dm³)	
	7-Day	28-Day
RC	2.35 ± 0.02	2.37 ± 0.01
RCF-25	2.33 ± 0.01	2.33 ± 0.00
RCF-50	2.30 ± 0.01	2.32 ± 0.00
RCF-50-ii	2.30 ± 0.01	2.31 ± 0.01
RC25	2.34 ± 0.01	2.35 ± 0.01
RC50	2.32 ± 0.01	2.33 ± 0.01
RC-50-ii	2.31 ± 0.01	2.32 ± 0.01
RC25-AS	2.32 ± 0.01	2.34 ± 0.01
RC50-AS	2.31 ± 0.01	2.33 ± 0.01
RC25-BR	2.33 ± 0.01	2.36 ± 0.02
RC50-BR	2.32 ± 0.01	2.34 ± 0.01

The table also shows that using mixed recycled aggregate (RA, RA_{WF}, RA_{AS} and RA_{BR}) yielded concretes with lower density than the reference concrete (RC), and that density declined with rising replacement ratios (Fig. 8). This finding also concurred with previous authors' reports [12] and [16].

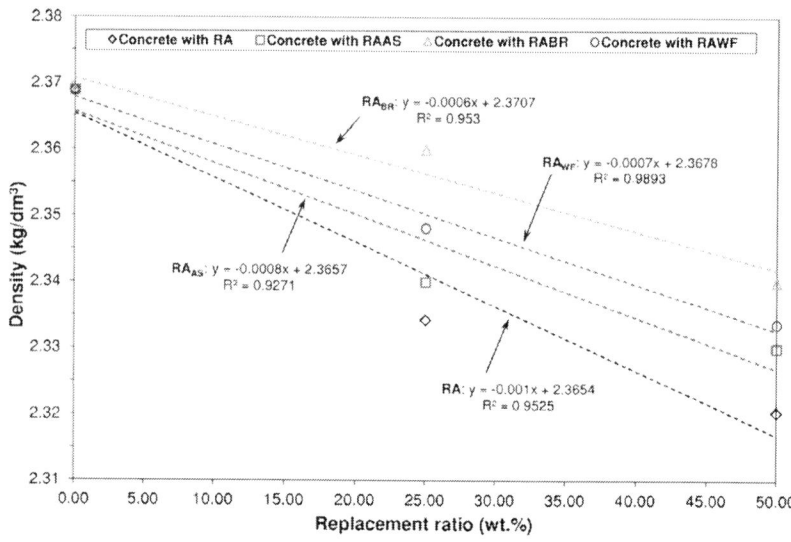

Figure 8: Density of the concretes at 28 days.

That decline was primarily the result of the lower density of the recycled than natural aggregate (see Table 5) and the effect of some of the recycled aggregate constituents (asphalt, brick and floating particles) on the aggregate/paste interface (see Section 3.2.6).

In addition, the recycled concretes RCF-50-ii and RC-50-ii at 28 days, prepared with RA_{II} and RA_{WFII}, respectively, were less dense than concretes RCF-50 and RC-50 (containing RA and RA_{WF}) at the same age, because the fines (<0.063 mm) of the C&DW have pozzolanic activity, as noted earlier [38].

As Fig. 8 shows and as described in earlier studies [6], the relationship between density and replacement ratio was linear.

Likewise according to the figure, independent of the replacement ratio, recycled concretes made with RA_{AS}, RA_{WF} and RA_{BR} were denser than the concretes containing mixed recycled aggregate (RA), a finding directly related to the higher density of the mixed recycled aggregates (RA_{AS}, RA_{WF} and RA_{BR}) (seeTable 5).

Compressive Strength of the Concrete Mixes

Fig. 9 depicts the variation in compressive strength of the concretes at 7 and 28 days; at the latter age, the mean strength was higher than 30 MPa and would be expected to continue to rise with age due to cement hydration [16].

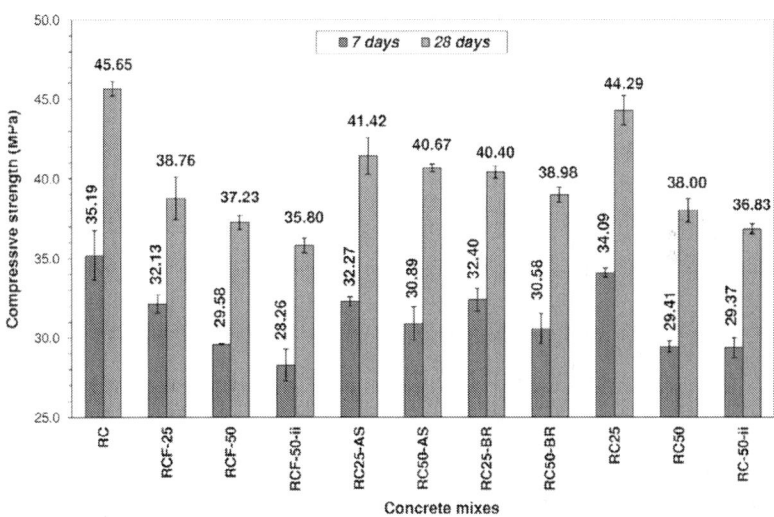

Figure 9: Compressive strength of the concrete mixes at 7 and 28 days.

The figure shows that the strength values declined with increasing replacement ratios at rates that depended on the constituents in the mixed recycled aggregate (RA, RA_{WF}, RA_{AS} y RA_{BR}), primarily because the ITZs between aggregate components and the paste (see Section 3.2.6) were weaker than the gravel/paste interfacial transition zone.

Further to the graph in Fig. 10, the relationship between replacement ratio and compressive strength at 28 days was clearly linear, and exhibited what is regarded as a high (>0.85) correlation coefficient [39].

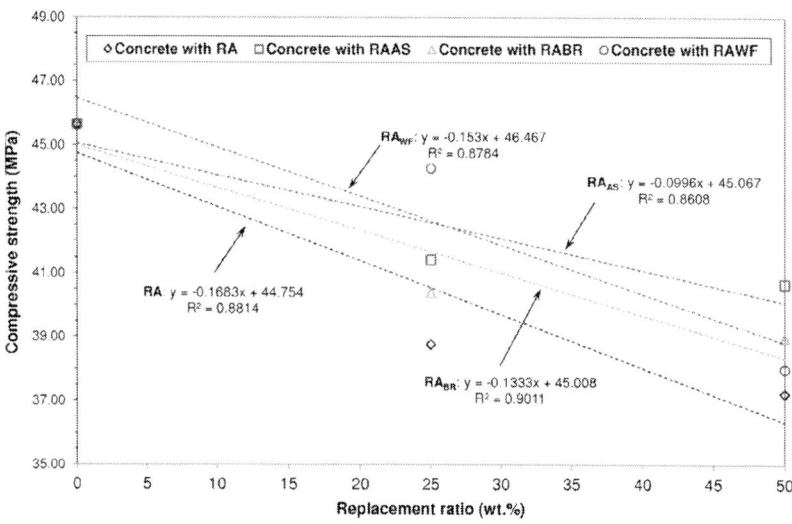

Figure 10: Compressive strength vs replacement ratio in concretes at 28 days.

The decline of the strength at 28 days with respect to the control (RC) for concretes prepared with the initial mixed recycled aggregate (RA) and the aggregate lacking asphalt (RA_{AS}), brick (RA_{BR}), floating particles (RA_{WF}), or both floating particles and fines (RA_{WFII}) was, respectively: 18.44%, 10.91%, 14.60%, 16.76%, 19.32% and 21.57%. Those findings were consistent with prior studies [6], [12], [16] and [10], some of which reported strength losses of 10–20% for RA replacement ratios of 25% and 50%.

Fig. 11 compares strength of the concretes RC25-AS, RC50-AS, RC25-BR, RC50-BR, RCF-50-ii, RC25, RC50 and RC50-ii at 7 and 28 days to strength at the same age in recycled concretes RCF-25 and RCF-50, prepared with the original mixed recycled aggregate (RA).

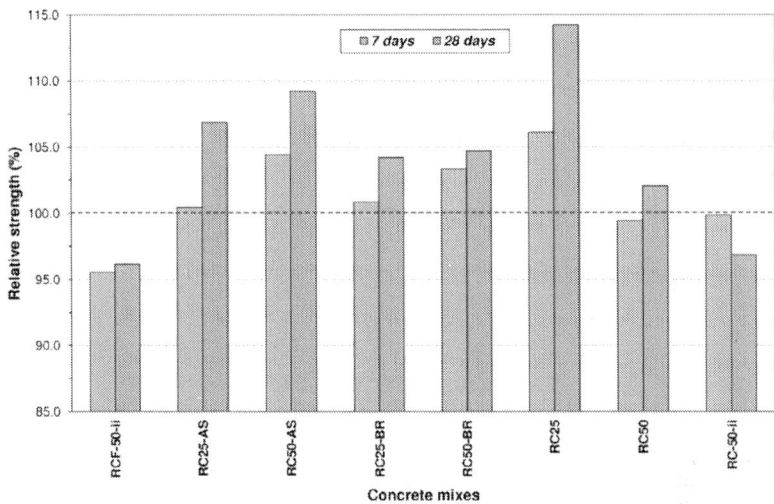

Figure 11: Relative strength of concrete mixes at 7 and 28 days (100% = RCF-25 and RCF-50).

The figure reveals the beneficial effect of removing the asphalt, brick and floating particles from the mixed recycled aggregate, which yielded strength values respectively 9.23%, 4.70% and 2.05% higher than in RCF-50. The greatest rise was observed when the asphalt was removed, for, as observed previously[40] and [41], the weak asphalt/paste interface determines poor bonding and consequently lowers the mechanical performance in concrete. The removal of the clay-based fraction induced a slight rise in strength. That is consistent with earlier reports [16], [10] and [42], where minor declines in strength were observed with the addition of different percentages (0–70%) of brick, roof tiles and similar, induced by the intrinsic properties of these materials [25].

The elimination of floating particles likewise induced a slight increase in strength due to the poor bonding between floating particles (wood and plastic) and the cement paste, which gave rise to a weak interfacial transition zone (ITZ). Similar findings can be found in the literature [15].

Lastly, removing fines (<0.063 mm) led to 3.85% and 3.17% lower strength in concretes containing RA_{WFII} and RA_{II} than in RCF-50 and RC-50, respectively. Such strength loss can be primarily attributed to the high clay-based (brick, roof tiles, sanitary ware) and mortar content of construction and demolition waste fines. Clay-based materials are characterised by a certain pozzolanic activity, which favours short- and long-term strength development [38], [43] and [44], whereas the attached mortar contains a certain percentage of anhydrous cement particles that can be rehydrated, thereby also enhancing strength [45].

Strength Indexes

The percentage contribution of the mixed recycled aggregate to 28 days concrete strength (P_{28}) was calculated in terms of strength indexes (see Table 7 and Fig. 12), found as per the methodology and nomenclature put forward in studies by Medina et al. [27] and Cachim [42]. The symbol q signifies the percentage of natural coarse aggregate, and R_b and R_r the contribution to strength respectively made by gravel and mixed recycled aggregate (RA, RA_{II}, RA_{WF}, RA_{WFII}, RA_{AS} and RA_{BR}) to strength. P represents the percentage contribution of mixed recycled aggregates to concrete strength. A P value higher than the replacement ratio (25% or 50%) means that the replacement aggregate contributed positively to strength.

Table 7: Strength indexes

Concrete mix	q	R_b	R_r	K	$P_{28días}$	$P_{7días}$
RC	100	0.46	0.00	1.00	0.00	0.00
RCF-25	75	0.52	0.06	1.13	11.69	17.86
RCF-50	50	0.74	0.29	1.63	38.70	40.52
RCF-50-ii	50	0.72	0.26	1.57	36.25	37.74
RC25	75	0.59	0.13	1.29	22.70	22.57
RC50	50	0.76	0.30	1.66	39.93	40.17
RC-50-ii	50	0.74	0.28	1.61	38.03	40.09
RC25-AS	75	0.55	0.10	1.21	17.35	18.21
RC50-AS	50	0.81	0.36	1.78	45.88	43.04
RC25-BR	75	0.52	0.07	1.15	12.77	18.54
RC50-BR	50	0.81	0.35	1.77	43.51	42.46

$$R_b = f_{ck}/q/R_r = R_b - R_{b(RC)}/P_{28} = 100 * (R_r/R_b).$$

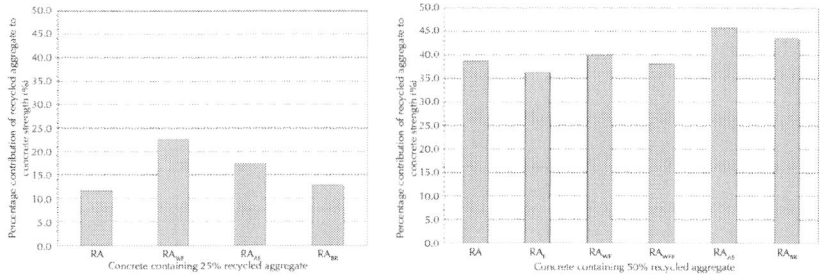

Figure 12: Percentage of the contribution of recycled aggregate to concrete strength at 28 days.

Table 7 lists the compressive strength index values for the concretes at 28 days, as well as the P values at 7 and 28 days. Further to the aforementioned definition, the replacement of natural by mixed recycled aggregate (RA, RA_{II}, RA_{WF}, RA_{WFII}, RA_{AS} and RA_{BR}) was observed to have an adverse effect ($P < 25$ or $P < 50$) on strength (Fig. 11).

These results (Fig. 12) provide further evidence that, irrespective of the replacement ratio, that effect was less intense when the asphalt, clay-based materials and floating particles were removed, and greater when the fines (<0.063 mm) were removed, regardless of the type of aggregate (R_{All} or RA_{WFII}).

Sorptivity

Table 8 gives the water sorptivity values obtained for the concretes. The sorptivity results appear to show that at a replacement ratio of 25%, sorptivity remained almost constant, with the recycled concrete mixes (i.e. RC25, RCF25, RC25-BR y RC25-AS) exhibiting marginally lower sorptivity than the reference concrete (RC), due to the effect of fine fraction that it is described in the next paragraph. At a replacement ratio of 50%, however, the recycled concrete mixes were found to have significantly higher sorptivity (15–46%) than the reference concrete.

Table 8: Concrete sorptivity

Concrete mix	S (mm/h$^{0.5}$)	Variation in sorptivity (%)
RC	0.684 ± 0.03[a]	0.00
RCF-25	0.668 ± 0.05	−2.34
RCF-50	0.788 ± 0.00	15.20
RCF-50-ii	1.001 ± 0.03	46.29
RC25	0.652 ± 0.06	−4.65
RC50	0.806 ± 0.03	17.86
RC-50-ii	0.991 ± 0.05	44.85
RC25-AS	0.648 ± 0.02	−5.32
RC50-AS	0.801 ± 0.02	17.16
RC25-BR	0.650 ± 0.03	−5.06
RC50-BR	0.804 ± 0.03	17.57

a± = Standard deviation.

The increase in sorptivity is consistent with the upward trend observed by other authors [46], [47],[48] and [49] using C&DW and ceramic waste [50] as coarse aggregate in concrete manufacture.

Table 8 also reveals the beneficial effect of fines on recycled concrete properties, for removing this fraction from the recycled aggregate induced higher sorptivity. These findings were due to the pozzolanic activity of this fraction, which by refining concrete porosity reduced the accessibility to water and other aggressive external agents. The effects of removing the other components (asphalt, clay-based materials and floating particles) were similar in all cases, due to the impact of mixed recycled aggregate water absorption (seeTable 5) on concrete water sorptivity, as reported by other authors [51].

Finally, further to a premise put forward in earlier papers [52] and [53], concrete durability can be ranked according to its water sorptivity. Under that premise, concretes with a water sorptivity of under 6 mm/h$^{0.5}$ are durable, although other authors [54] have proposed lowering that value to 3 mm/h$^{0.5}$ for reasons of safety. Based on those criteria, the recycled concretes made with this type of recycled aggregate may be regarded as durable for housing applications.

ITZs of Carse Aggregates/paste

The morphology of the ITZs between the recycled aggregate (concrete waste, asphalt, clay-based materials (sanitary ware and tile) and floating particles) and the paste is shown in the micrographs in Fig. 13.

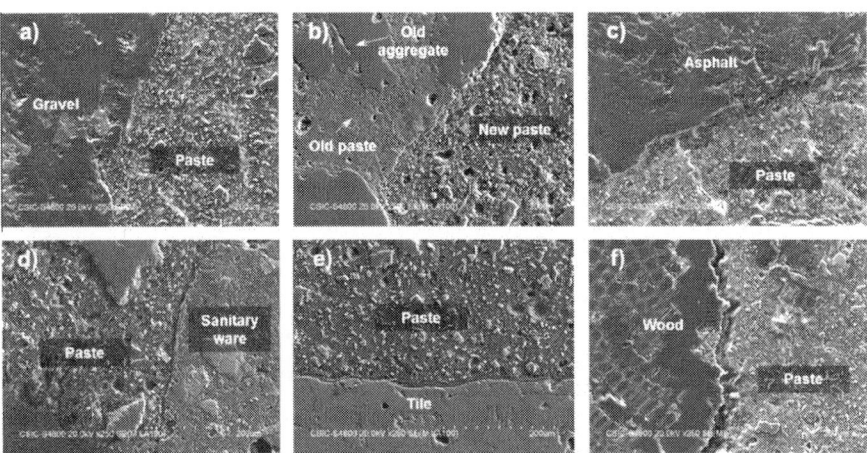

Figure 13: Coarse aggregate/paste ITZ (×250): (a) gravel; (b) concrete waste; (c) asphalt; (d) and (e) clays-based materials; and (f) floating particles (wood).

This figure shows that the natural aggregate, concrete waste and clay-based materials (sanitary ware and tile) formed a less porous, more compact and more continuous interface than the organic components (asphalt and wood), due to the poor bonding between the latter and cement paste, as observed in earlier studies [41] and [55].

The variation in the Ca/Si ratio in the region extending radially outward from the surface of the aggregate, which gradually penetrates the cement paste (ITZs), is shown in Fig. 14. The ITZ can be regarded to consist of four phases: pores, calcium hydroxide, other hydration products (primarily calcium hydrate and ettringite) and unhydrated cement grains [56].

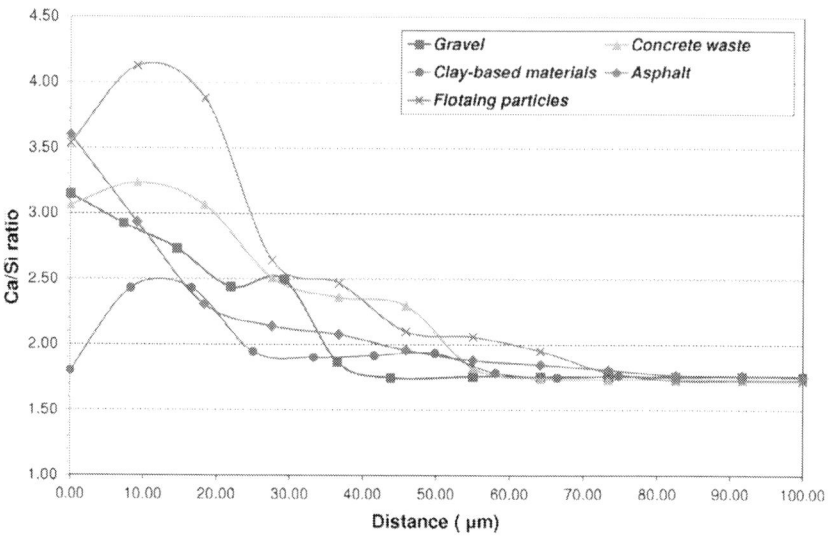

Figure 14: Variation in Ca/Si ratio in ITZs.

As the figure shows, the Ca/Si ratio declined gradually along these ITZs to practically the same value in all cases: 1.76, characteristic of calcium silicate hydrates (C–S–H gel). This value lies within the ranges proposed by other authors [1.2–2.3 [57]; and 0.8–2.5 [58]], and is similar to 1.6, the ratio found earlier by the present team in recycled concretes [34].

Finally, Fig. 14 shows that the thickness of the ITZ varied depending on aggregate type, ranging from 35 to 50 μm in gravel/paste, 50 to 60 μm in concrete and clay-based waste/paste and 65–80 μm in asphalt and wood/paste. The values for the natural aggregate/paste and concrete waste/paste ITZs were in line with the literature: 10–50 μm [59] and [60] for the former, and 30–65 μm [61], [62] and [63] for the latter.

CONCLUSIONS

The conclusions that can be drawn from the present findings and observations are listed below:

- The saturated density and mechanical performance of recycled concretes were found to be moderately lower than in the reference concrete, particularly at the higher mixed recycled aggregate replacement ratio.

- The remove of asphalt, clay-based materials and floating particles yielded higher compressive strength concretes than the original mixed recycled aggregate (RA). By contrast, the exclusion of fines from the recycled aggregate induced a decline in recycled concrete performance.

- Water sorptivity in recycled concrete decreased with 25 wt% replacement ratios and rose when the ratio was 50 wt%.

- The coarse aggregate/paste interface varied depending on the components, with inorganic materials (gravel, concrete waste and clay-based materials) exhibiting a narrower and more compact interface than organic constituents (asphalt and floating particles).

- Based on the mechanical performance and water sorptivity found for recycled concretes, these materials would be apt for use in housing applications.

- Satisfactory construction and demolition waste management delivers good quality recycled aggregate which in turn yields higher performing recycled concretes.

ACKNOWLEDGMENTS

This research has been made possible through funding from the Spanish Ministry for Science and Innovation's national projects ref. BIA 2010-21194-C03-01, BIA 2013-48876-C3-1-R and BIA 2013-48876-C3-2-R. The main part of the experimental study was carried out in the UK collaboratively with the University of the West of Scotland, during the research visit by the first author.

REFERENCES

1.   European Commission. Service contract on management of construction and demolition waste – SR1. Available from <http://ec.europa.eu/environment/waste/pdf/2011_CDW_Report.pdf18/07/2013>; 2011.

2.   Belin P, Habert G, Thiery M, Roussel N. Cement paste content and water absorption of recycled concrete coarse aggregates. Mater Struct 2013:1–15.

3.   Coelho A, de Brito J. Economic viability analysis of a construction and demolition waste recycling plant in Portugal – part I: location, materials, technology and economic analysis. J Clean Prod 2013;39:338–52.

4.   British Standard Institution. BS 8500-2. Concrete – Complementary British Standard to BS EN 206-1. Part 2: specification for constituent materials and concrete; 2006a.

5.   Vegas I, Ibanez JA, Lisbona A, Saez de Cortazar A, Frias M. Pre-normative research on the use of mixed recycled aggregates in unbound road sections. Constr Build Mater 2011;25(5):2674–82.

6.   Martínez-Lage I, Martínez-Abella F, Vázquez-Herrero C, Pérez-Ordóñez JL. Properties of plain concrete made with mixed recycled coarse aggregate. Constr Build Mater 2012;37:171–6.

7.   Kwan WH, Ramli M, Kam KJ, Sulieman MZ. Influence of the amount of recycled coarse aggregate in concrete design and durability properties. Constr Build Mater 2012;26(1):565–73.

8.   Etxeberria M, Mari AR, Vazquez E. Recycled aggregate concrete as structural material. Mater Struct 2007;40(5):529–41.

9.   Limbachiya MC. Recycled aggregates: production, properties and value-added sustainable applications. J Wuhan Univ Technol Mater Sci Ed 2010;25(6):1011–6.

10.   Yang J, Du Q, Bao Y. Concrete with recycled concrete aggregate and crushed clay bricks. Constr Build Mater 2011;25(4):1935–45.

11.   Mas B, Cladera A, Bestard J, Muntaner D, López CE, Piña S, et al. Concrete with mixed recycled aggregates: influence of the type of cement. Constr Build Mater 2012;34:430–41.

12.   Mas B, Cladera A, Olmo TD, Pitarch F. Influence of the amount of mixed recycled aggregates on the properties of concrete for non-structural use. Constr Build Mater 2012;27(1):612–22.

13.   Agrela F, Sánchez de Juan M, Ayuso J, Geraldes VL, Jiménez JR. Limiting properties in the characterisation of mixed recycled aggregates for use in the manufacture of concrete. Constr Build Mater 2011;25(10):3950–5.

14. Park W, Noguchi T. Influence of metal impurity on recycled aggregate concrete and inspection method for aluminum impurity. Constr Build Mater 2013;40:1174–83.

15. Ferreira L, de Brito J, Saikia N. Influence of curing conditions on the mechanical performance of concrete containing recycled plastic aggregate. Constr Build Mater 2012;36:196–204.

16. Sheen Y-N, Wang H-Y, Juang Y-P, Le D-H. Assessment on the engineering properties of ready-mixed concrete using recycled aggregates. Constr Build Mater 2013;45:298–305.

17. British Standard Institution. BS EN 197-1. Cement. Composition, specifications and conformity criteria for common cements; 2011.

18. Teychenné DC, Franklin RE, Erntroy HC. Design of normal concrete mixes. Garston, Watford – United Kingdom: IHS BRE Press; 2010.

19. British Standard Institution. BS 8500:1. Concrete–Complementary British Standard to BS EN 206-1. Part 1: method of specifying and guidance for the specifier; 2006b.

20. British Standard Institution. BS EN 12620. Aggregates for concrete; 2008.

21. German Standard Institution. DIN 4226-100. Aggregates for mortar and concrete. Part 100. Recycled aggregates; 2002.

22. Laboratório Nacional de Engenharia Civil. Especificação LNEC E – 471: Guia para a utilização de agregados reciclados grossos em betões de gigantes hidráulicos; 2006.

23. European Committee for Standardization. CEN/TC 104/SC 1/TG 19 – Use of aggregates in concrete; 2013. [Draft].

24. de Juan MS, Gutierrez PA. Study on the influence of attached mortar content on the properties of recycled concrete aggregate. Constr Build Mater 2009;23(2):872–7.

25. Paine KA, Dhir RK. Recycled aggregates in concrete: a performance-related approach. Mag Concr Res 2010;62(7):519–30.

26. Mansur MA, Wee TH, Cheran LS. Crushed bricks as coarse aggregate for concrete. ACI Mater J 1999;96(4):478–84.

27. Medina C, Frías M, Sánchez de Rojas MI, Thomas C, Polanco JA. Gas permeability in concrete containing recycled ceramic

sanitary ware aggregate. Constr Build Mater 2012;37:597–605.

28.   Limbachiya MC, Marrocchino E, Koulouris A. Chemical-mineralogical characterisation of coarse recycled concrete aggregate. Waste Manage 2007;27(2):201–8.

29.   Angulo SC, Ulsen C, John VM, Kahn H, Cincotto MA. Chemical-mineralogical characterization of C&D waste recycled aggregates from Sao Paulo, Brazil. Waste Manage 2009;29(2):721–30.

30.   Bianchini G, Marrocchino E, Tassinari R, Vaccaro C. Recycling of construction and demolition waste materials: a chemical-mineralogical appraisal. Waste Manage 2005;25(2):149–59.

31.   Vegas I, Ibanez JA, Jose JTS, Urzelai A. Construction demolition wastes, Waelz slag and MSWI bottom ash: a comparative technical analysis as material for road construction. Waste Manage 2008;28(3):565–74.

32.   Rodrigues F, Carvalho MT, Evangelista L, de Brito J. Physical–chemical and mineralogical characterization of fine aggregates from construction and demolition waste recycling plants. J Clean Prod 2013;52:438–45.

33.   Gonzalez-Fonteboa B, Martinez-Abella F. Concretes with aggregates from demolition waste and silica fume. Materials and mechanical properties. Build Environ 2008;43(4):429–37.

34.   Medina C, Frías M, Sánchez de Rojas MI. Microstructure and properties of recycled concretes using ceramic sanitary ware industry waste as coarse aggregate. Constr Build Mater 2012;21:112–8.

35.   Senthamarai RM, Manoharan PD, Gobinath D. Concrete made from ceramic industry waste: durability properties. Constr Build Mater 2011;25(5):2413–9.

36.   Lin J, Chen M, Wu S. Utilization of silicone maintenance materials to improve the moisture sensitivity of asphalt mixtures. Constr Build Mater 2012;33:1–6.

37.   Comisión Permanente del Hormigón. Instrucción Hormigón Estructural. EHE- 08. Ministerio de Fomento. Centro de Publicaciones, Madrid; 2008.

38.   Asencio, E., Frías, M., Sánchez de Rojas, M.I., and Medina, C. Design of new cement matrixes based on construction and demolition waste. 1st International & 3rd National Congress on

Construction Sustainable and EcoEfficient\ Solutions, Sevilla (Spain); 2013.

39. de Brito J, Alves F. Concrete with recycled aggregates: the Portuguese experimental research. Mater Struct 2010;43:35–51.

40. Huang BS, Shu X, Li GQ. Laboratory investigation of Portland cement concrete containing recycled asphalt pavements. Cem Concr Res 2005;35(10):2008–13.

41. Huang B, Shu X, Burdette EG. Mechanical properties of concrete containing recycled asphalt pavements. Mag Concr Res 2006;58(5):313–20.

42. Cachim PB. Mechanical properties of brick aggregate concrete. Constr Build Mater 2009;23(3):1292–7.

43. Lin KL, Wu HH, Shie JL, Hwang CL, Cheng A. Recycling waste brick from construction and demolition of buildings as pozzolanic materials. Waste Manage Res 2010;28(7):653–9.

44. Sánchez de Rojas MI, Marin F, Rivera J, Frias M. Morphology and properties in blended cements with ceramic wastes as a pozzolanic material. J Am Ceram Soc 2006;89(12):3701–5.

45. Kim YJ, Choi YW. Utilization of waste concrete powder as a substitution material for cement. Constr Build Mater 2012;30:500–4.

46. Kou SC, Poon CS. Enhancing the durability properties of concrete prepared with coarse recycled aggregate. Constr Build Mater 2012;35:69–76.

47. Wirquin E, Hadjieva-Zaharieva R, Buyle-Bodin F. Use of water absorption by concrete as a criterion of the durability of concrete – application to recycled aggregate concrete. Mater Struct 2000;33(230):403–8.

48. Olorunsogo FT, Padayachee N. Performance of recycled aggregate concrete monitored by durability indexes. Cem Concr Res 2002;32(2):179–85.

49. Katz A. Properties of concrete made with recycled aggregate from partially hydrated old concrete. Cem Concr Res 2003;33(5):703–11.

50. Medina C, Sanchez de Rojas MI, Frias M. Properties of recycled ceramic aggregate concretes: water resistance. Cement Concr Compos 2013;40:21–9.

51. Zega CJ, Di Maio ÁA. Use of recycled fine aggregate in concretes with durable requirements. Waste Manage 2011;31(11):2336–40.

52. Alexander M, Ballim Y, Stanish K. A framework for use of durability indexes in performance-based design and specifications for reinforced concrete structures. Mater Struct 2008;41(5):921–36.

53. Ho DWS, Hinczak I, Conroy JJ, Lewis, RK. Influence of slag cement on the water sorptivity of concrete. Proc. Fly ash, silica fume, slag and natural puzzolans in Concrete. ACI SP 91-72, Madrid, Spain; 1986. p. 1463–1473.

54. Menendez G, Bonavetti VL, Irassar EF. Ternary blend cement concrete. Part II: transport mechanisms. Mater Constr 2007;57(285):31–43.

55. Etxeberria M, Vazquez E, Mari A. Microstructure analysis of hardened recycled aggregate concrete. Mag Concr Res 2006;58(10):683–90.

56. Li W, Xiao J, Sun Z, Kawashima S, Shah SP. Interfacial transition zones in recycled aggregate concrete with different mixing approaches. Constr Build Mater 2012;35:1045–55.

57. Hewlett PC. Lea's chemistry of cement and concrete, London; 1998.

58. Erdem S, Dawson AR, Thom NH. Influence of the micro- and nanoscale local mechanical properties of the interfacial transition zone on impact behavior of concrete made with different aggregates. Cem Concr Res 2012;42(2):447–58.

59. Lee KM, Park JH. A numerical model for elastic modulus of concrete considering interfacial transition zone. Cem Concr Res 2008;38(3):396–402.

60. Zheng JJ, Li CQ, Zhou XZ. Thickness of interfacial transition zone and cement content profiles around aggregates. Mag Concr Res 2005;57(7):397–406.

61. Li J, Xiao H, Zhou Y. Influence of coating recycled aggregate surface with pozzolanic powder on properties of recycled aggregate concrete. Constr Build Mater 2009;23(3):1287–91.

62. Poon CS, Shui ZH, Lam L. Effect of microstructure of ITZ on compressive strength of concrete prepared with recycled aggregates. Constr Build Mater 2004;18(6):461–8.

63. Xiao J, Li W, Sun Z, Lange DA, Shah SP. Properties of interfacial transition zones in recycled aggregate concrete tested by nanoindentation. Cement Concr Compos 2013;37:276–92.

Citations

CHAPTER 1

Abdelrahman Osman Elfaki, Saleh Alatawi, and Eyad Abushandi, "Using Intelligent Techniques in Construction Project Cost Estimation: 10-Year Survey," Advances in Civil Engineering, vol. 2014, Article ID 107926, 11 pages, 2014. doi:10.1155/2014/107926.

CHAPTER 2

S. Silva, A. Araújo, D. Costa and J. L. Meliá, "Safety Climates in Construction Industry: Understanding the Role of Construction Sites and Workgroups," Open Journal of Safety Science and Technology, Vol. 3 No. 4, 2013, pp. 80-86. doi: 10.4236/ojsst.2013.34010.

CHAPTER 3

Awad, R. and Banerjee, S. (2014) Some Construction Methods of A-Optimum Chemical Balance Weighing Designs.Journal of Applied Mathematics and Physics, 2, 1159-1170. doi: 10.4236/jamp.2014.213136.

CHAPTER 4

Karim A AbdelWarith, Panagiotis Ch Anastasopoulos, Wayne Richardson, Jon D Fricker, and John E Haddock, Design of local roadway infrastructure to service sustainable energy facilities, doi:10.1186/2192-0567-4-14.

CHAPTER 5

André Torre, Romain Melot, Habibullah Magsi, Luc Bossuet, Anne Cadoret, Armelle Caron, Ségolène Darly, Philippe Jeanneaux, Thierry Kirat, Haï Vu Pham, and Orestes Kolokouris, Identifying and measuring land-use and proximity conflicts: methods and identification, doi:10.1186/2193-1801-3-85.

CHAPTER 6

Stuart J Self, Bale V Reddy, and Marc A Rosen, Review of Underground Coal Gasification Technologies and Carbon Capture, doi: 10.1186/2251-6832-3-16.

CHAPTER 7

Gianluca Corrias, Roberta Licheri, Roberto Orrù, Giacomo Cao, Optimization of the self-propagating high-temperature process for the fabrication in situ of Lunar construction materials, Chemical Engineering Journal, Volumes 193–194, 15 June 2012, Pages 410-421, ISSN 1385-8947, http://dx.doi.org/10.1016/j.cej.2012.04.032.

CHAPTER 8

C. Medina, W. Zhu, T. Howind, M. Frías, M.I. Sánchez de Rojas, Effect of the constituents (asphalt, clay materials, floating particles and fines) of construction and demolition waste on the properties of recycled concretes, Construction and Building Materials, Volume 79, 15 March 2015, Pages 22-33, ISSN 0950-0618, http://dx.doi.org/10.1016/j.conbuildmat.2014.12.070.

Index